ANTIGEN-PRESENTING CELLS

IN FOCUS

Titles published in the series:

*Antigen-presenting Cells

*Complement

Enzyme Kinetics

Gene Structure and Transcription

Genetic Engineering

*Immune Recognition

*Lymphokines

Membrane Structure and Function

Molecular Basis of Inherited Disease

Regulation of Enzyme Activity

*Published in association with the British Society for Immunology.

Series editors

David Rickwood
Department of Biology, University of Essex, Wivenhoe Park, Colchester, Essex CO4 3SQ, UK

David Male
Institute of Psychiatry, De Crespigny Park, Denmark Hill, London SE5 8AF, UK

ANTIGEN-PRESENTING CELLS

Jonathan M.Austyn

Nuffield Department of Surgery, University of Oxford,
John Radcliffe Hospital, Headington, Oxford OX3 9DU, UK

at
OXFORD UNIVERSITY PRESS
Oxford New York Tokyo

IRL Press
Eynsham
Oxford
England

© IRL Press at Oxford University Press 1989

First published 1989

All rights reserved by the publisher. No part of this book may be reproduced or transmitted in any form by any means, electronic or mechanical, including photocopying, recording or any information storage and retrieval system, without permission in writing from the publisher.

British Library Cataloguing in Publication Data

Austyn, J.M.
　Antigen-presenting cells
　1. Man. Immune reactions
　I. Title.　II. Series
　616.07'95

ISBN 0 19 963005 4

Library of Congress Cataloging-in-Publication Data

Austyn, Jonathan M.
　Antigen-presenting cells / J.M.Austyn.
　　p.　cm. — (In focus)
　Includes bibliographies and index.
　ISBN 0-19-963005-4 (soft): $11.95 (U.S.)
　1. Antigen presenting cells.　I. Title.　II. Series: In focus
　(Oxford, England)
　[DNLM: 1. Antigen-Presenting Cells—immunology.　QW 568 A958a]
　QB185.8.A59A97 1989
　612'.118222—dc19

Typeset and printed by Information Press Ltd, Oxford, England.

Preface

The function of lymphocytes in the immune system is to recognize foreign antigens and directly or indirectly to bring about their removal from the body. T lymphocytes recognize antigens bound to major histocompatibility complex (MHC) molecules on the surface of cells called antigen-presenting cells (APCs). For some time, APCs were thought of as a particular set of cells expressing MHC class II molecules. Two main ideas to emerge over the past few years necessitate a re-evaluation of this concept. First, it now seems that any cell in the body may be able to present antigens to T cells of one type or another; the T cell then secretes antigen-non-specific mediators to potentiate the function of the APC or to kill it. Second, there is evidence that a specialized subset of antigen-presenting leukocytes is required to activate resting T cells, a process that may be termed immunostimulation. Together, APCs and immunostimulatory cells can be referred to as accessory cells.

Although there are many unknowns, we are gaining insights into the mechanisms whereby samples of a cell's internal contents are 'processed', bound to MHC molecules, and presented at the cell surface for perusal by T lymphocytes; this property provides a crucial link between the immune system and the body in general. We are also beginning to understand the process of immunostimulation. Moreover, it is becoming clear that the migration of a distinct lineage of dendritic leukocytes through the body, and their various developmental stages, play pivotal roles in determining how, when, and where immune responses are induced *in vivo*. Such knowledge is crucial to our understanding of the immune system, and opens exciting possibilities for immunological manipulations. Many of these ideas are yet to be found in current texts; this book is offered as a modest attempt to redress this situation.

J.M.Austyn

Acknowledgements

I am grateful to all who permitted me to use published and unpublished material in this book, including my graduate students Chris Larsen and Paul Fairchild. Ralph Steinman's helpful suggestions as to its organization are also gratefully acknowledged. I thank Gilla Kaplan for the loan of her PC in New York, and Myron and Sylvia Alpert for their kindness and the opportunity to be inspired by the colours of a late fall in Pennsylvania while writing.

Contents

Abbreviations	ix

1. Immune recognition

Introduction	1
B Cells and Antibodies	1
T Cells, Lymphokines and Cytolysins	3
Lymphocyte Activation	6
Antigen Recognition by T Cells	6
MHC restriction of T cells	6
T cell recognition of processed antigens	11
Peptide–MHC interactions	12
Self-restriction and alloreactivity	13
General Texts	15
Further Reading	15
References	16

2. Antigen presentation in immune responses

Introduction	17
Antigen Presentation by Macrophages	17
Macrophage activation	17
Interferon-γ in macrophage activation	20
Delayed-type hypersensitivity	21
Immune Rejection of Allografts	21
Interferon-γ in allogeneic reactions	23
Antigen Presentation in B Cell Responses	23
Two Types of Helper T Cell	24
Further Reading	26
References	26

3. Accessory cells in culture

	28
Introduction	28
Immunostimulatory Cells: Dendritic Cells	28
Features	28

Functions	32
Antigen-presenting Cells	35
Dendritic Leukocytes *In Situ*	36
Interdigitating cells of T areas	37
Follicular dendritic cells of B areas	37
Veiled cells of afferent lymph	39
Langerhans cells of skin	41
Dendritic leukocytes in non-lymphoid organs	41
Summary	43
Further Reading	43
References	44

4. Antigen processing and immunostimulation

Introduction	46
Antigen Processing and Presentation	46
Macrophages	47
B cells	49
Fibroblasts	51
Mechanisms of Immunostimulation	54
T cell activation	54
Clustering	54
Lymphocyte activating signals	55
Antigen presentation by dendritic leukocytes	56
Further Reading	57
References	57

5. Accessory cells *in vivo*

Dendritic Cell Function *In Vivo*	59
Migration and Maturation of Dendritic Leukocytes	60
Immunostimulation in lymph nodes	61
Dendritic Leukocytes in Transplantation	62
Stimulation of allograft rejection	62
Central versus peripheral sensitization	64
Dendritic Cells in the Thymus	67
T cell development	67
The restriction repertoire	69
Self-tolerance	71
Further Reading	72
References	72
Glossary	74
Index	77

Abbreviations

ADCC	antibody-dependent cell-mediated cytotoxicity
Ag	antigen
AIDS	acquired immunodeficiency syndrome
APC	antigen-presenting cell
ATXBM	adult thymectomized X-irradiated bone marrow-reconstituted
BCG	Bacillus Calmette Guerin
CD	cluster of differentiation
Con A	concanavalin A
CR3	complement receptor type 3
CSF	colony stimulating factor
CTL	cytotoxic lymphocyte
CURL	compartment of uncoupling of receptors and ligands
DC	dendritic cell
dGuo	deoxyguanosine
DNA	deoxyribonucleic acid
DTH	delayed-type hypersensitivity
GM-CSF	granulocyte–macrophage CSF
ER	endoplasmic reticulum
FDC	follicular DC
FITC	fluorescein isothiocyanate
H-2	histocompatibility locus 2 (mouse MHC)
HA	haemagglutinin
HEV	high endothelial venules
HIV	human immunodeficiency virus
HLA	human leukocyte antigen (human MHC)
HRP	horseradish peroxidase
Ia	class II MHC molecules
IDC	interdigitating cell
IFNγ	interferon-γ
Ig	immunoglobulin
IL	interleukin
kd	kilodalton
KLH	keyhole limpet haemocyanin
LC	Langerhans cell
LFA	lymphocyte function-associated antigen

LPS	lipopolysaccharide
β_2-m	β_2-microglobulin
Mϕ	macrophage
MHC	major histocompatibility complex
mIg	membrane Ig
MLR	mixed leukocyte reaction
mRNA	messenger RNA
NP	nucleoprotein
PMA	phorbol myristate acetate
RER	rough ER
RNA	ribonucleic acid
T_C, T_H, T_S	cytotoxic, helper and suppressor T cell subsets
TCR	T cell antigen receptor
TD/TI	T-dependent/T-independent
TDL	thoracic duct lymphocytes
TMV	tobacco mosaic virus
TNF	tumour necrosis factor
TNP	trinitrophenyl
VC	veiled cell

Amino acids (three-letter code)

Ala	alanine
Arg	arginine
Asp	aspartate
Gln	glutamine
Glu	glutamate
Gly	glycine
Ile	isoleucine
Lys	lysine
Leu	leucine
Met	methionine
Phe	phenylalanine
Ser	serine
Thr	threonine
Tyr	tyrosine
Val	valine

1
Immune recognition

1. Introduction

The immune system is responsible for destroying things that are foreign to the body, including many pathogenic organisms, tumour cells, and allografts—that is tissues transplanted from genetically different individuals. Recognition of foreign antigens is mediated by lymphocytes, B cells and T cells, which have distinct though structurally related types of cell surface antigen receptors, antibodies (immunoglobulins; Igs) and T cell receptors (TCRs) (*Figure 1.1*). Each lymphocyte has multiple copies of one particular receptor on its plasma membrane so that it recognizes and responds to just one antigen or antigenic determinant. The immense diversity of immune responses is made possible by having an enormous number of lymphocytes with different specificities. It has been estimated that there are some 2×10^{12} lymphocytes in the human body, similar in total mass to the liver or brain, and the total number or repertoire of antigen receptor specificities that can potentially be generated from the germ-line exceeds even this number; in the case of antibodies it is diversified still further through somatic mutation. While B cells and T cells are similar, in that both recognize foreign antigens, their functions are very different, mainly because they secrete different types of molecules during immune responses.

2. B cells and antibodies

The antigen receptors of B cells, membrane Ig, are produced in a secreted form when these cells develop into plasma cells. Secretory antibodies represent one of the body's effector mechanisms which may be soluble, as in this case, or cellular in nature but all bring about or 'effect' removal of antigens from the body. Antibodies bind to free or soluble antigens, such as viruses and bacterial toxins, and mediate their destruction. For example the complement cascade, a series of plasma proteins, can be activated on antibodies bound to an organism causing holes to be produced in its surface, thus allowing the entry of water, small ions and polyelectrolytes into the cell and hence its destruction by osmotic and colloidal forces (necrosis). Other activated complement components recruit and stimulate inflammatory cells, particularly phagocytes such as neutrophils

2 Antigen-presenting cells

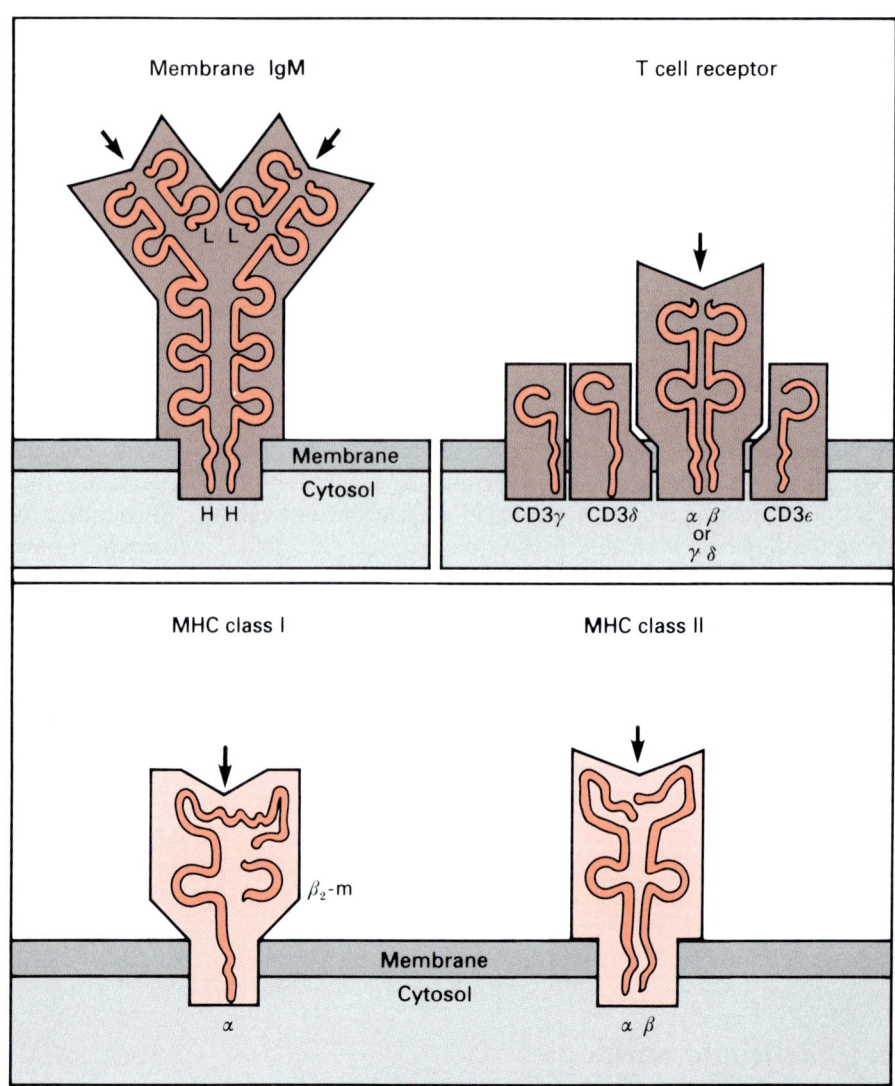

Figure 1.1. Antigen-binding molecules. Antibodies such as membrane IgM and the T cell antigen receptor are members of the immunoglobulin superfamily. Their polypeptide chains are folded into one or more domains (loops) each containing a characteristic structure, the Ig fold, as well as intrachain disulphide bonds. The approximate sites of antigen binding are indicated (arrows). MHC molecules which bind antigenic peptides are also members of this family. (Based on ref. 30.) β_2-m, β_2-microglobulin.

and monocytes. Particles to which complement components and/or antibodies are bound are 'opsonized', meaning they are more readily internalized and digested by phagocytes, uptake being mediated by membrane receptors such as complement receptors and Fc receptors which bind the Fc portions of some antibodies.

3. T cells, lymphokines and cytolysins

In contrast to B cells, T cells recognize foreign antigens on the surface of cells of the body; examples include viral antigens on the surface of virus-infected cells and transplantation antigens on allografts. Specific recognition is mediated by TCRs which can be one of two types ($\alpha\beta$ or $\gamma\delta$ heterodimers) but additional membrane components, often called ancillary molecules, are involved in T cell antigen recognition and functions (*Table 1.1*). T cells are divided functionally into cytotoxic, helper and suppressor subsets. Cytotoxic T (T_C) cells are effector cells that kill target cells expressing foreign antigens. Helper T (T_H) cells are regulatory cells that control the development and function of effector cells, while suppressor T (T_S) cells inhibit the function of helper and/or effector cells and will not be considered further.

One of the remarkable things about the immune system and the focus of this book, is the ability of T cells to recognize infected, aberrant or otherwise 'non-self' cells that *contain* foreign organisms and other antigens, whereas antibody produced by B cells recognizes free antigens. It is probable that any cell in the body can display representative samples of its intracellular contents at the plasma membrane for perusal by T cells. Cells that 'present' foreign antigens to T cells are called antigen-presenting cells, and the process is referred to as antigen presentation.

The function of T cells is to recognize foreign antigens and produce antigen-non-specific mediators at the surface of the antigen-presenting cell (APC). Helper T cells secrete lymphokines that help the APC develop into, or become better, effector cells, whereas T_C cells produce cytolysins which kill the APC. It makes biological sense for cells not to express a large number of defences constitutively but for these to be induced in effector cells when they are required, or for the cells to be destroyed. Thus T_H cells stimulate B cells to develop into

Table 1.1. T cell ancillary molecules

Molecule	Nomenclature of homologues in: human	mouse	rat	Some known or possible functions
CD2	T11			Binds LFA-3[a]
				Can mediate T cell activation
CD3	T3			Associated with TCR
				Mediates signal transduction
CD4	T4	L3T4	W3/25	Binds MHC class II; mediates adhesion
				(? indirectly) and signal transduction
CD8	T8	Ly2	OX8	Binds MHC class I; mediates adhesion
				(? indirectly) and signal transduction
CD11a/CD18[b]	LFA-1[a]	LFA-1		Binds ICAM-1 and other ligands
				Mediates adhesion

[a]LFA—lymphocyte function-associated antigen.
[b]CD11a is the α chain of LFA-1; CD18 is the β chain, shared with CR3 and p150,95.

Table 1.2. T cell-derived lymphokines

Lymphokines[a]	Synonyms	Examples of effects on different lineages				
		T cells	B cells[b]	Mononuclear phagocytes	Polymorphs	Others
IL2	TCGF	Activated T cell growth, lymphokine release; cytotoxic activity	Proliferation; Ig-secretion	Cytotoxicity in some populations		NK activity increased
IL3	multi-CSF	Supports growth of some T cell lines	Supports growth of pre-B lines		Eosinophil CSF	Stem cell growth, differentiation; mast cell growth
IL4	BCGF-I BSF1	Growth of T cells and thymocytes; cytotoxicity	Antigen expression; Ig-secretion (e.g. IgE, IgG1)	Giant cell formation; tumoricidal activity		Mast cell growth
IL5	BCGF-II	Cytotoxic T cells from thymocytes (with IL2)	Proliferation; Ig-secretion (e.g. IgA)		Eosinophil differentiation	
IL6	BSF2	Proliferation of thymocytes (with PMA & IL4)	Growth of plasmacytomas; Ig-secretion			Hepatic cells: production of acute phase proteins
IFNγ	MAF		Proliferation; Ig-secretion (e.g. IgG2a)	Macrophage activation		MHC induction

[a] Some differences between mouse and human.
[b] Details in Chapter 2.
Blank boxes do not necessarily imply that the lymphokine has no effect on those cells. CSF, colony stimulating factor; NK, natural killer; PMA, phobol myristate acetate.

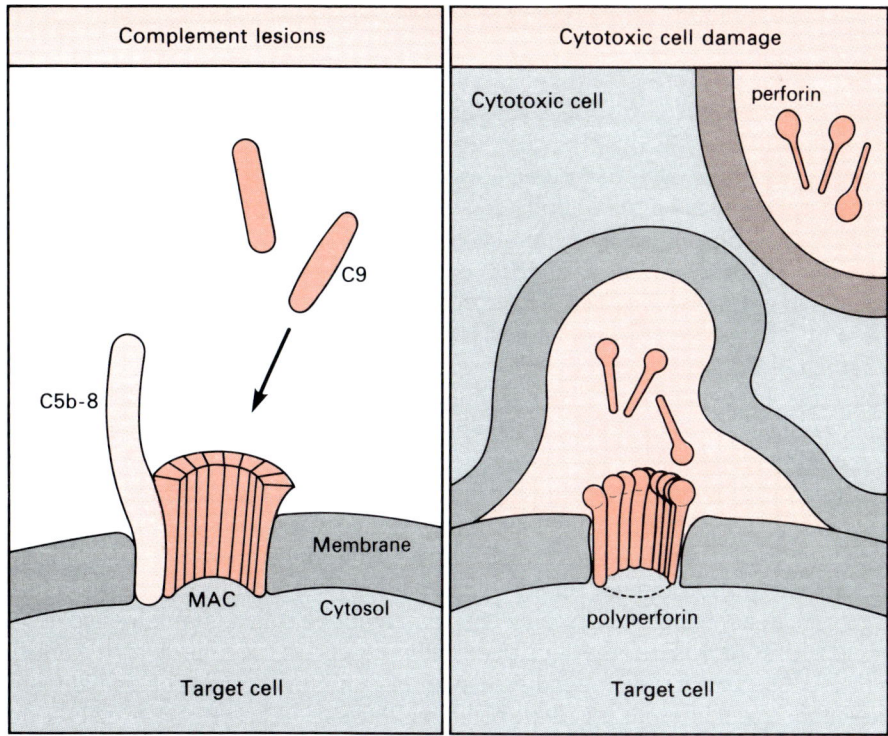

Figure 1.2. Target cell damage by immunological effector systems. Both humoral and cell-mediated damage can be caused to target cells. C5b-8, inserted into the membrane of the target cell, induces the polymerization of C9 to form the membrane attack complex. (MAC). Cytotoxic cells can release perforin from their granules in close apposition to the target cell membrane which polymerize to form pores on the target cell. Perforin is structurally similar to C9 and forms similar lesions to the MAC.

antibody-secreting plasma cells, and enable macrophages to acquire more effective microbicidal and cytocidal capabilities, whereas T_C cells presumably eliminate cells when they are no longer able to develop adequate defences.

Lymphokines include molecules designated according to the interleukin (IL) nomenclature, for example IL2 through IL6, and others such as interferon-γ (IFNγ) (*Table 1.2*). (Strictly speaking the term 'lymphokine' refers to a soluble molecule made by any lymphocyte, not just a T cell; 'interleukin' is used for certain molecules that 'communicate between leukocytes'.) Cytolysins include perforins which form pores in target cell membranes and lead to destruction of the cell by osmotic/colloidal forces, somewhat analogous to the action of later complement components to which they are related (*Figure 1.2*). These pores may facilitate the entry of non-pore-forming cytolysins such as tumour necrosis factor(s) (TNFβ and TNFα also made by macrophages), serine esterases, and molecules that trigger DNA degradation and initiate an 'auto-suicide' mechanism for target cell death (apoptosis).

4. Lymphocyte activation

In the absence of foreign antigens, lymphocytes are small, resting cells in G_0 phase of the cell cycle. Resting T cells are essentially inactive and cannot secrete lymphokines or cytolysins, but they do express TCRs and ancillary molecules on their membrane. Before antigen-specific T cells (and lymphocytes in general) can function they need to be activated and moved into the cell cycle (*Figure 1.3*). Entry to G_1 is accompanied by an increase in size or 'blastogenesis', increased RNA and protein synthesis, and the expression of new membrane molecules (activation antigens) required for subsequent functions. This is followed by a phase of DNA synthesis, and mitosis when cell division occurs.

Activated or 'sensitized' T cells respond to APCs in two ways. First, they proliferate to generate a larger population of specific effector cells (clonal expansion) as well as memory cells which produce a faster and more efficient 'secondary' response should the antigen be encountered again. Second, they secrete antigen-non-specific mediators at the surface of the APC and either kill it or help it develop into a better effector cell.

It follows that there are two critical events in a T cell response, both requiring antigen recognition. Small, resting lymphocytes first need to be activated into large, functional lymphoblasts, an event that might be called immunostimulation (or 'sensitization'). Then the sensitized cells respond in the appropriate manner to the APCs. There is now good evidence that immunostimulation and antigen presentation are different processes (Chapter 3). Moreover, specialized cells seem to be required to activate resting T cells, whereas any cell in the body may become an APC for sensitized T cells, and we use the term APC in the latter sense. Cells that express foreign antigens and stimulate responses of resting or sensitized lymphocytes can be collectively termed accessory cells. The rest of this chapter is devoted to an outline of how T cells recognize foreign antigens.

5. Antigen recognition by T cells

5.1 MHC restriction of T cells

Specialized membrane molecules act as 'guidance systems' and enable T cells to recognize antigens on cell surfaces; these are MHC (major histocompatibility complex) molecules (1). MHC genes were discovered in early studies on the survival of skin and tumour allografts. A number of histocompatibility loci were found to control the tempo of allograft rejection in mice, and of these the most important (major) genes were grouped together on one chromosome as a complex, which for historical reasons is called histocompatibility locus 2 (H-2) in the mouse. The corresponding set of genes in the human is called the human leukocyte antigen (HLA), and an MHC has been identified in all mammals examined. Allograft rejection was later shown to be primarily a (T) cell-mediated response (2), and it turns out that MHC genes control rejection in these experimental or clinical situations by virtue of their physiological role in permitting antigen

Immune recognition 7

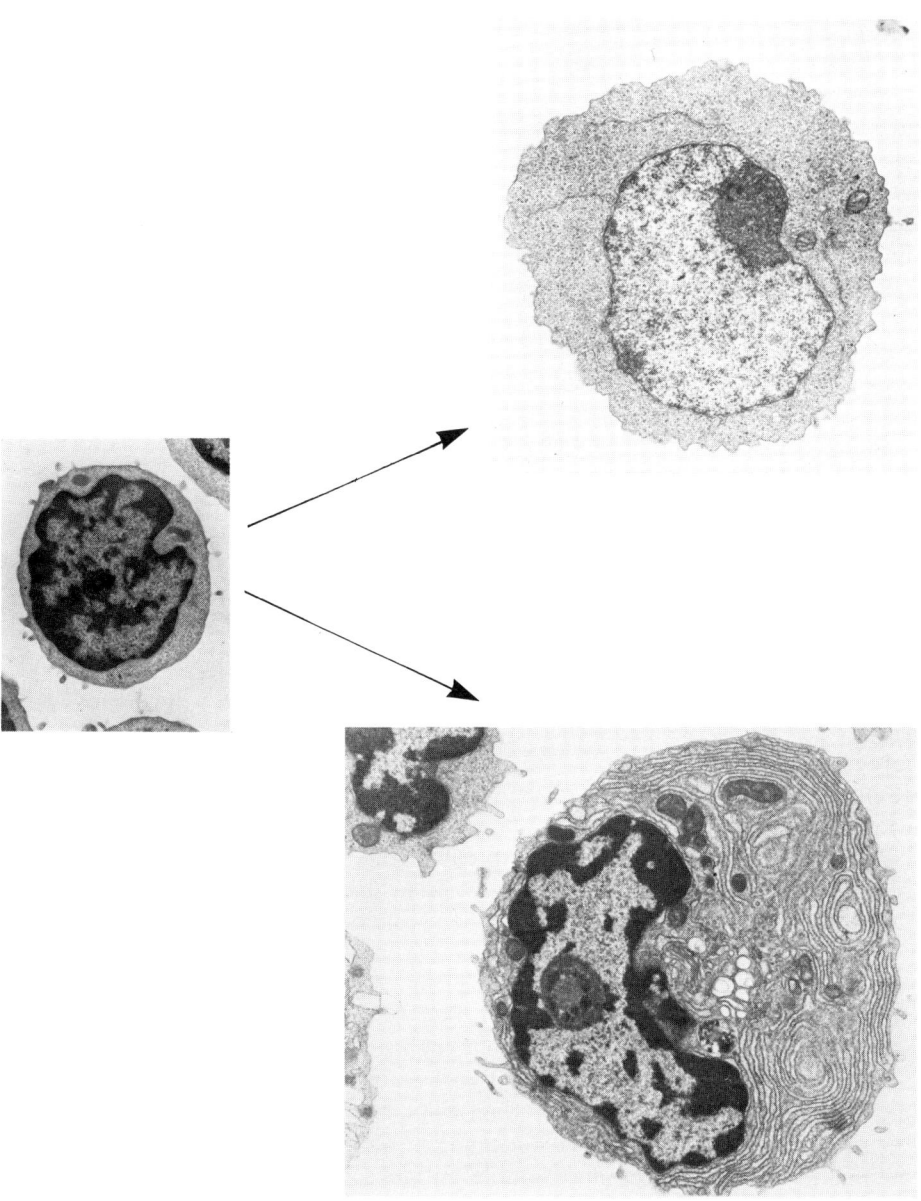

Figure 1.3. Lymphocyte activation. Small resting B cells or T cells (which appear similar in electron micrographs: left) need to be activated (right) before then can participate in immune responses. Resting T cells develop into T cell blasts (top right) which divide and secrete a variety of lymphokines, while activated B cells can develop into antibody-secreting plasma cells (bottom right). Note the increase in cytoplasmic:nuclear ratio on activation (top right) and the extensive rough endoplasmic reticulum in plasma cells (bottom right) which is required for antibody production often at rates up to 2000 molecules sec^{-1}. Electron micrographs courtesy of Professor C.Grossi and Professor A.Zicca.

recognition by T cells.

There are two classes of MHC genes and molecules. Class I MHC molecules are expressed on essentially all cells of the body (3), although normally their levels vary from weak or almost undetectable (e.g. many endocrine cells and corneal endothelium) to very strong (e.g. T cells). Expression of these molecules can be up-regulated on some cell types during immune responses (4). Class II MHC molecules (or Ia molecules; the terms are interchangeable) have a more limited distribution and are constitutively expressed on very few cell types, notably lymphoid dendritic cells and B cells. They are inducible on macrophages and levels can vary from almost undetectable to very high. Class II molecules can also be induced in the course of immune responses on other cell types that may not normally express them, such as certain epithelia (4–7). Nevertheless, class II molecules have a more limited tissue distribution than class I. (Note that cells expressing Ia molecules are often called 'APCs' but we use this term in a more general and now more accurate sense: see Section 4.)

MHC molecules are heterodimers (see *Figure 1.1*). Class I molecules are composed of a 45 kilodalton (kd) α chain encoded in the MHC, together with a 12 kd non-MHC product called β_2 microglobulin (β_2-m). Both the α and β chains of class II molecules are encoded by the MHC and their masses range from about 27 to 35 kd. In the mouse the main class I loci are H-2K and H-2D, while the class II loci are found in the H-2I region which is divided into IA and IE subregions (*Figure 1.4*). In humans, the respective class I loci are called HLA-A, -B and -C, while the class II loci are located in the HLA-D region which is subdivided into at least DP, DQ and DR subregions (*Figure 1.4*). Each class II subregion contains one or more α and β chains. The polypeptide chains of MHC molecules are folded into a number of extracellular domains, each containing a characteristic structure called the 'Ig fold' (from which the Ig superfamily derives its name—see *Figure 1.1*). The class I α chain has three such domains, while the class II α and β chains have two each.

MHC molecules on cell surfaces have two main functions. First, they bind 'foreign antigens', particularly as peptide fragments. Second, the MHC molecule and/or foreign peptide may then be recognized by a specific TCR and, in this way, MHC molecules allow T cells to recognize cell-associated antigens. Since T cells are *restricted* to recognizing antigens bound to (or 'associated with') MHC molecules rather than antigens on their own, this phenomenon is termed MHC restriction or MHC-associative recognition. In contrast, B lymphocytes and antibodies recognize free and soluble antigens that are not bound to MHC molecules (although there are a few reports of apparently MHC-restricted antibodies).

In general, different types of T cell recognize antigens bound to different classes of MHC molecule. Many T_C cells and most T_H cells recognize antigens bound, respectively, to MHC class I and class II molecules. This correlation and the different distributions of class I and class II molecules suggests that any cell in the body may need to become a target for (class I-restricted) T_C cells, for example after virus infection, but that some cells are better able to become effector cells and eliminate the antigen after they are recognized by (class

Immune recognition 9

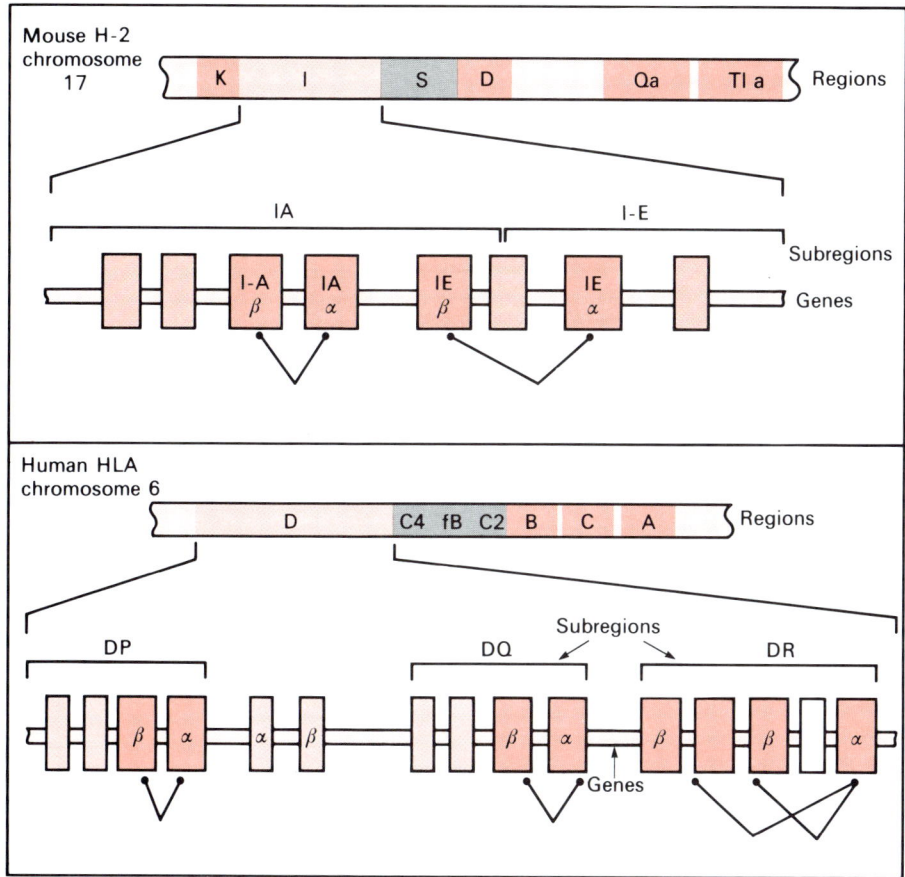

Figure 1.4. Mouse and human MHC. There are three classes of MHC genes. Class I genes (deep orange) encode molecules which can be expressed on virtually all cells of the body and are very polymorphic. In mouse the majority of class I molecules are encoded by the Qa and Tla loci and these are less polymorphic and have a more limited tissue distribution. Class II genes (medium orange) encode molecules that are expressed by only some cells of the body. The class II loci are shown expanded beneath the H-2 and HLA maps. Pseudogenes are lighter in colour; blocks without colour represent genes that may or may not be pseudogenes, or which are present in only some haplotypes. Linked pairs in the class II regions indicate known associations of heterodimers. Class III genes (grey) encode complement components, some cytokines and some enzymes.

II-restricted) T_H cells. These are approximations, however, for some T_C cells are class II-restricted and some T_H cells are class I-restricted.

When attempts were made to distinguish functionally different T cells according to phenotype, it was found that many helper cells expressed the CD4 molecule, while many T_C cells carried CD8 (*Table 1.1*). These molecules are reciprocally expressed on two subsets of mature T cells in the blood and lymphoid tissues (i.e. $CD4^+CD8^-$ or $CD4^-CD8^+$), while a large number of immature

10 Antigen-presenting cells

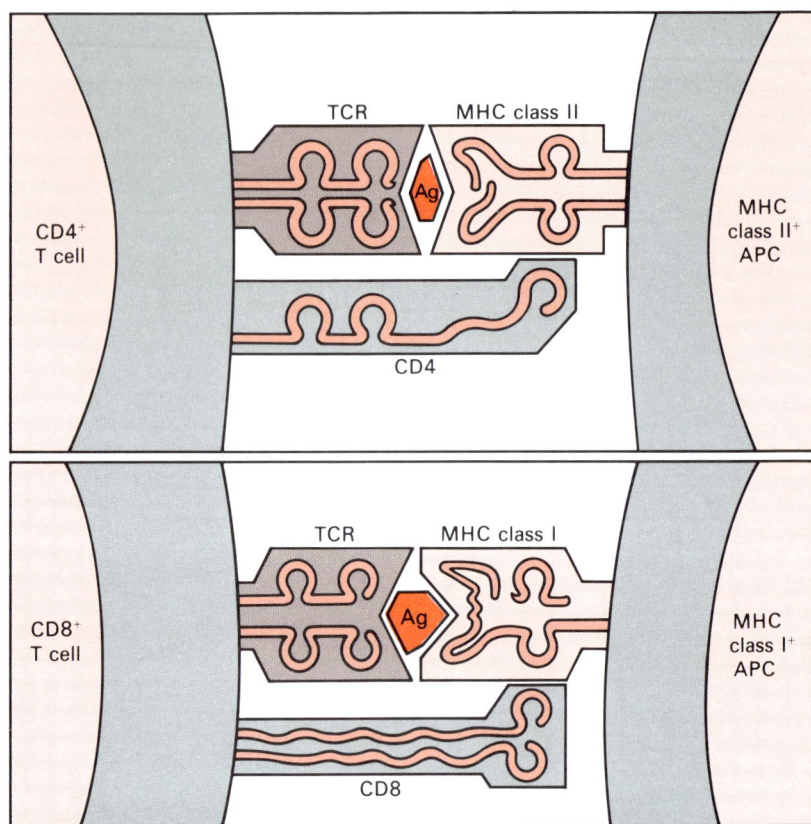

Figure 1.5. Role of CD4 and CD8 in immune recognition. CD4 and CD8 are members of the Ig superfamily which interact, respectively, with MHC class II and class I molecules on APC, mediating T cell adhesion (perhaps by recruiting other adhesion systems), and signal transduction to the T cell. The interaction sites on the MHC molecules are not known with certainty; the lengths of the CD8 polypeptide chains shown here are exaggerated relative to CD4. Many T_H cells express CD4 while most T_C cells express CD8.

T cells in the thymus are 'double-positive' and express both molecules ($CD4^+CD8^+$) and the most immature are double-negative.

Possession of CD4 or CD8 was found to correlate more closely with the cells' ability to recognize antigens bound, respectively, to class II or class I MHC molecules, rather than their function (*Figure 1.5*). Even here there are a few exceptions, and currently there are no phenotypic markers that correlate precisely with T cell function (8,9). Cytotoxic and T_H cells can use common pools of TCR V genes, so it is unlikely these impart class I or class II restriction on the cell. It appears that CD4 and CD8 play important roles in this, probably by binding to the respective class of MHC molecule, mediating adhesion between T cells and APCs (10,11), and perhaps also delivering stimulatory or inhibitory signals to the T cell.

5.2 T cell recognition of processed antigens

Some 30 years ago, it was realized that humoral- and cell-mediated responses were often made against different forms of antigen. Guinea pigs were immunized with proteins and their response to challenge with the same protein in its native (e.g. naturally folded) or denatured conformations was examined (12). Anaphylaxis (a form of hypersensitivity mediated by antibodies, also called immediate or type I hypersensitivity) was generated primarily by the native molecule, whereas delayed-type hypersensitivity responses in the skin (mediated by T cells and also called type IV hypersensitivity) were produced with either native or denatured antigen. It was later found that antibodies produced against globular proteins preferentially bound to the native molecule, whereas T cells responded in culture to both native and denatured antigens in the presence of APCs. Furthermore, antibodies against native antigens failed to inhibit T cell responses, while anti-MHC antibodies did inhibit (13). The idea to emerge was that antibodies tended to recognize 'conformational' epitopes exposed in the three-dimensional structure of the antigen, while T cells seemed to recognize 'sequential' epitopes that were revealed when the molecule was unfolded or degraded (*Figure 1.6*).

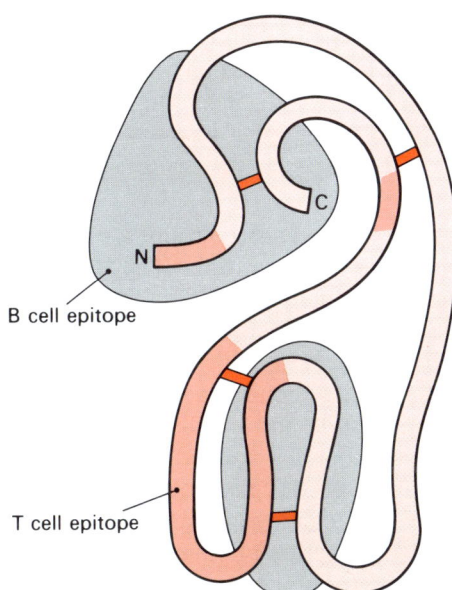

Figure 1.6. T and B cell epitopes. Some T and B cell (Ig) epitopes are shown for the globular protein hen egg lysozyme (deep orange indicates disulphide bonds). While T cells often recognize linear amino acid sequences within the protein (medium orange), antibodies recognize regions on the surface of the molecule that are not necessarily juxtaposed in a linear sequence (grey backshaded areas) (31).

The phenomenon whereby native antigen is converted to an altered or 'processed' form that can be recognized by T cells became known as antigen processing. Cloned T_H cells, for example $CD4^+CD8^-$ T cells that produced helper lymphokines, were found to respond both to native antigens and to certain fragments of antigen, for example peptides, in the presence of APCs (14). Thus APCs processed antigens, and T_H cells recognized peptides (plus MHC). However, it was believed that T_C cells recognized antigens, such as viral proteins, inserted into the membrane of APCs as intact molecules rather than fragments.

It is now clear that helper and cytotoxic T cells are similar in their ability to recognize processed antigens, bound to MHC molecules on cell surfaces. A central observation was that virus-specific cytotoxic lymphocytes (CTL) killed virus-infected cells just as well as uninfected target cells that were incubated with small synthetic peptides (15). This work was facilitated by the definition of T cell epitopes in molecular terms, thus allowing pure antigenic peptides to be synthesized. By testing the response of T cell clones and lines to fragments of natural antigen, then to 'nested sets' of synthetic peptides (overlapping in sequence but becoming progressively shorter), and ultimately to peptides with defined amino acid substitutions, T cell epitopes have been mapped with considerable precision. Most range from about 9 to 25 residues in length, and are often buried within the native conformation. Retrospectively, attempts were made to predict where T cell epitopes occur in proteins, first (16), by examining the molecule for regions that theoretically might be able to adopt the structure of amphipathic α-helix (with hydrophobic and hydrophilic residues on opposite sides), and second (17), by use of a consensus sequence (particularly a sequence starting with a charged residue or glycine, followed by two or three consecutive hydrophobic residues, and terminating in a polar amino acid).

5.3 Peptide–MHC interactions

Since T cells recognize processed antigens and are MHC-restricted, binding of antigen to MHC, or this combination to the TCR in a ternary complex, would be expected. A variety of techniques including equilibrium dialysis (18) and gel filtration (19) has been used successfully to show binding of certain antigenic peptides to MHC class II molecules. Other data indicated that the presence of a specific TCR was sometimes required to stabilize an inherently low affinity interaction between the peptide and MHC (20).

In some studies, certain peptides seemed to bind specifically to MHC class II molecules of one haplotype but not another, apparently correlating with whether or not the MHC molecule was from an individual that responded to the peptide (e.g. 18,19). The concept that immune responsiveness is determined by the MHC (also known as 'determinant selection') is part of current dogma, even though the interactions are often quite weak and close analysis of the data indicates that specificity is not always absolute. Moreover, there are instances in which peptides have a relatively high affinity for class II molecules, but are non-immunogenic in that haplotype (21).

There is a report (22) that peptides can bind directly to MHC class I molecules, and a number of studies indicate that competition can occur between peptides for binding to an apparently single site, or closely overlapping sites, on class I (and more recently class II) molecules (23,24). The response of class I-restricted CTLs to a particular peptide/MHC combination has been found to be inhibited by a different peptide when they are added together during the cytotoxicity assay, or when target cells are pre-treated with both peptides and washed before adding CTLs (indicating that competition occurs on the APC). For reasons which are not clear, it was not possible to compete by pre-treating APCs with either peptide alone. Competing peptides were sometimes closely related in structure, or completely unrelated and even restricted through different MHC molecules. The physiological relevance of this, especially in a possible excess of potentially competing self-peptides, is uncertain.

The idea of a single binding site for peptides on MHC molecules is compatible with the structure of a class I molecule revealed by X-ray crystallography (25). The molecule contains a 'groove' between two α-helices sitting on a relatively flat β-pleated sheet, formed by association of the two membrane-distal domains $\alpha 1$ and $\alpha 2$. This 'sausages-on-a-grill' structure is supported by the membrane-proximal $\alpha 3$ domain. The groove is where peptides are thought to be bound, especially as it is lined by polymorphic residues that determine T cell responsiveness (see Section 5.4) and some unexplained electron density was found in this region and interpreted as bound peptide(s) (*Figure 1.7*). The groove is large enough to accommodate molecules of the size of known T cell epitopes in an extended conformation (e.g. ~8 residues) or contracted in an α-helix (e.g. ~20 residues).

5.4 Self-restriction and alloreactivity

There is evidence that in physiological situations T cells preferentially recognize foreign antigens/peptides bound to self-MHC molecules on APCs, that is 'self-restriction'. In immunology, 'self' is defined as that environment in which the immune system develops, thus phrased because it is possible for lymphocytes to develop in a genetically different environment, for example after allogeneic bone marrow transplantation when T cells originating from the donor marrow develop in a different thymus. Such T cells 'learn', or more correctly are selected, to recognize preferentially antigens bound to MHC molecules of the recipient rather than their own MHC, but are unresponsive or tolerant to the recipient's MHC and other molecules in the absence of foreign antigen (26).

MHC molecules are highly polymorphic (i.e. there are multiple alleles at each locus, at least 50 in some cases, in the population as a whole, although any heterozygote of course possesses only two of each). A high proportion of T cells from any individual can respond to MHC molecules from MHC-disparate (allogeneic) members of the species. This phenomenon of alloreactivity was first recognized in studies on graft-versus-host responses (27); these can be induced, e.g. by injecting T cells from one individual (e.g. strain C) into an F1 animal (e.g. strain C × D): the recipient cannot reject C strain T cells because it is

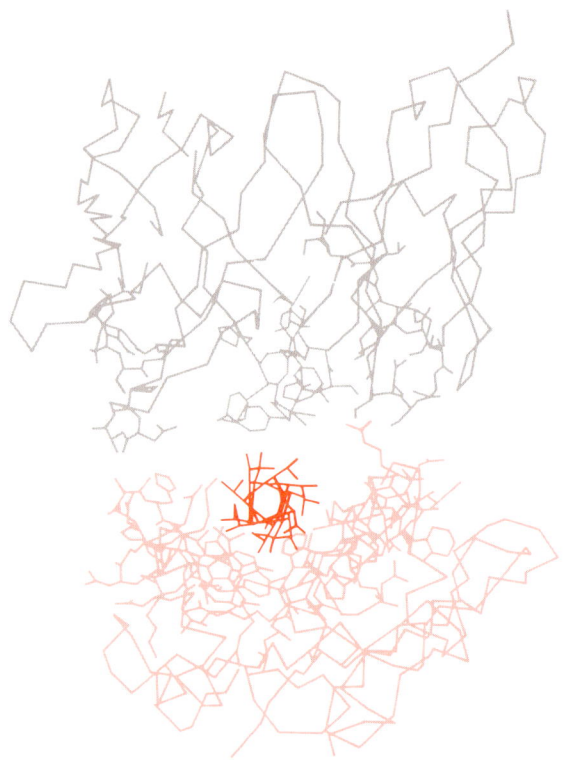

Figure 1.7. Antigen-binding sites. Representation of the structure of a hypothetical TCR (tan) interacting with an antigenic peptide (dark orange) held in α-helical configuration on the cleft of an HLA-A2 molecule (orange) as seen from the side. Redrawn from ref. 32.

tolerant, but the latter react against D strain allo-antigens. Alloreactivity has been studied in culture (28) in the mixed leukocyte reaction (MLR) in which T cells proliferate when they are incubated with allogeneic leukocytes.

It has been estimated that between 0.1 and 0.5% of the T cells in any individual can respond to a particular class I difference, while 1–10% of the T cell population responds to a fully allogeneic MHC. These figures and the number of alleles in the population mentioned above indicates there is likely to be overlap in the populations of T cells responding to different MHC molecules. One explanation for alloreactivity is that many T cells can use the same receptor to recognize foreign antigens bound to self-MHC, and an allogeneic MHC molecule possibly associated with a different peptide (29) (*Figure 1.8*). The phenomenon is of major importance because it may be the most potent stimulus for graft rejection.

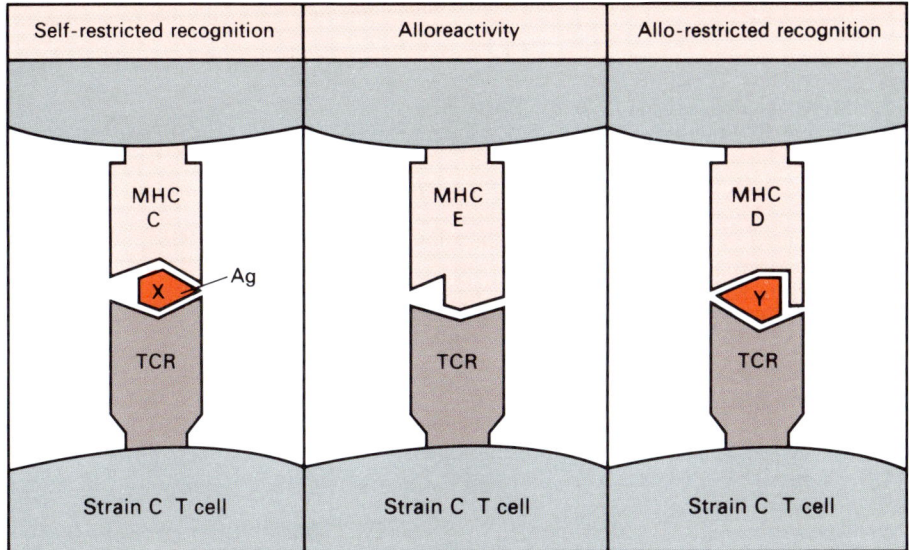

Figure 1.8. Recognition of antigen/MHC. The diagram illustrates how a TCR may recognize one antigen (X) bound to self-MHC (C), allogeneic MHC (E), or another antigen (Y) presented on a different allogeneic MHC molecule (D). Such cross-reactive recognition provides the basis for one explanation of alloreactivity, that is alloreactive cells are actually specific for antigen plus self-MHC in physiological situations.

6. General texts

Benjamini,E. and Leskowitz,S. (1987) *Immunology: A Short Course*. Alan R.Liss, New York/Wiley, Chichester.
Golub,E.S. (1987) *Immunology: A Synthesis*. Sinauer/Blackwell Scientific, Oxford.
Clark,W.R. (1986) *The Experimental Foundations of Modern Immunology*. John Wiley & Sons, Chichester.
Male,D., Champion,B. and Cooke,A. (1987) *Advanced Immunology*. Gower, London/Lippincott, Philadelphia.
Alberts,B., Bray,D., Lewis,J., Raff,M., Roberts,K. and Watson,J.D. (1983) *Molecular Biology of the Cell*. Garland, New York.
Watson,J.D., Tooze,J. and Kurtz,D.T. (1983) *Recombinant DNA: A Short Course*. Freeman, New York.

7. Further reading

Owen,M.J. and Lamb,J.R. (1988) *Immune Recognition* (In Focus series). IRL Press, Oxford.
Hamblin,A.S. (1988) *Lymphokines* (In Focus series). IRL Press, Oxford.
Law,S.K.A. and Reid,K.B.M. (1988) *Complement* (In Focus series). IRL Press, Oxford.
Davis,M.M. and Bjorkman,P.J. (1988) T-cell antigen receptor genes and T-cell recognition. *Nature*, **334**, 395.

8. References

1. Klein,J. (1986) *Natural History of the Major Histocompatibility Complex.* John Wiley & Sons, New York.
2. Medawar,P.B. (1944) *J. Anat.,* **78**, 176.
3. Daar,A.S., Fuggle,S.V., Fabre,J.W., Ting,A. and Morris,P.J. (1984) *Transplantation,* **38**, 287.
4. Milton,A.D. and Fabre,J.W. (1985) *J. Exp. Med.,* **161**, 98.
5. Barclay,A.N. and Mason,D.W. (1982) *J. Exp. Med.,* **156**, 1665.
6. Lampert,I.A., Suitters,A.J. and Chisholm,P.M. (1981) *Nature,* **293**, 149.
7. Mason,D.W., Dallman,M. and Barclay,A.N. (1981) *Nature,* **293**, 150.
8. Schwartz,R.H. (1985) *Annu. Rev. Immunol.,* **3**, 237.
9. Swain,S.L. (1983) *Immunol. Rev.,* **74**, 129.
10. Doyle,C. and Strominger,J.L. (1987) *Nature,* **330**, 256.
11. Norment,A.M., Salter,R.D., Parham,P., Englehard,V.H. and Littman,D.R. (1988) *Nature,* **336**, 79.
12. Gell,P.G.H. and Benacerraf,B. (1959) *Immunology,* **2**, 64.
13. Ellner,J.J., Lipsky,P.E. and Rosenthal,A.S. (1977) *J. Immunol.,* **118**, 2053.
14. Thomas,D.W., Hseieh,K.-H., Schauster,J.L. and Wilner,G.D. (1981) *J. Exp. Med.,* **153**, 583.
15. Townsend,A.R.M., Rothbard,J., Gotch,F.M., Bahadur,G., Wraith,D. and McMichael,A.J. (1986) *Cell,* **44**, 959.
16. Rothbard,J.B. (1986) *Ann. Inst. Pasteur,* **137**, 518.
17. DeLisi,C. and Berzofsky,J.A. (1985) *Proc. Natl. Acad. Sci. USA,* **82**, 7048.
18. Babbitt,B., Allen,P., Matsueda,G., Haber,E. and Unanue,E. (1985) *Nature,* **317**, 359.
19. Buus,S., Sette,A., Colon,S.M., Miles,C. and Grey,H.M. (1987) *Science,* **235**, 1353.
20. Watts,T.H., Gaub,H.E. and McConnell,H.M. (1986) *Nature,* **320**, 179.
21. Guillet,J., Lai,M., Briner,T.J., Buus,S., Sette,A., Grey,H.M., Smith,J.A. and Gefter,M.L. (1987) *Science,* **235**, 865.
22. Chen,B.P. and Parham,P. (1989) *Nature,* **337**, 743.
23. Maryanski,J.L., Pala,P., Cerrotini,J. and Corradin,G. (1988) *J. Exp. Med.,* **167**, 1391.
24. Bodmer,H., Bastin,J., Askonas,B. and Townsend,A. (1989) *Immunology,* **66**, 163.
25. Bjorkman,P.J., Saper,M.A., Samraoui,B., Bennett,W.S., Strominger,J.L. and Wiley,D.C. (1987) *Nature,* **329**, 512.
26. Schwartz,R.H. (1984) In *Fundamental Immunology.* Paul,W.E. (ed.), Raven Press, New York.
27. Ford,W.L., Simmonds,S.J. and Atkins,R.C. (1975) *J. Exp. Med.,* **141**, 681.
28. Wilson,D.B., Blyth,H. and Nowell,P.C. (1968) *J. Exp. Med.,* **128**, 1157.
29. Matis,L.A., Sorger,S.B., McElligott,D.L., Fink,P.J. and Hedrick,S.M. (1987) *Cell,* **51**, 59.
30. Williams,A.F. and Barclay,A.N. (1988) *Annu. Rev. Immunol.,* **6**, 381.
31. Male,D., Champion,B. and Cooke,A. (1987) *Advanced Immunology.* Gower, London/Lippincott, Philadelphia, p. 8.2.
32. Davis,M.M. and Bjorkman,P.J. (1988) *Nature,* **334**, 395.

Antigen presentation in immune responses

1. Introduction

Antigen recognition by lymphocytes is a prerequisite for all adaptive immune responses. Recognition occurs during the afferent phase when resting T cells are activated (immunostimulation) and during the efferent or effector phase of the response when the sensitized T cells may encounter the antigen on a variety of antigen-presenting cells (APCs). Helper T (T_H) cells are particularly important in controlling the immune response because they can recognize antigens presented by cells like macrophages and B cells, and then release lymphokines that enable these cells, as well as cytotoxic T cell (T_C) precursors, to develop into potent effector cells (*Figure 2.1*).

Unlike antibodies, T cell antigen receptors (TCRs) are not subject to somatic hypermutation. This implies an individual T cell can potentially recognize the same antigen/major histocompatibility complex (MHC) structure during its development and in both the afferent and efferent phases of the immune response. It is important to differentiate between these three distinct stages when T cell recognition may occur. T cell development occurs in the thymus where the T cell repertoire and self-tolerance are determined (at least for helper cells), whereas immunostimulation of resting T cells occurs mainly in secondary lymphoid tissues: these events are further discussed in Chapter 5. The role of antigen presentation in the effector phase of the immune response, which occurs especially in the periphery, is outlined in this chapter.

2. Antigen presentation by macrophages

2.1 Macrophage activation

Macrophages are mononuclear phagocytes. This lineage includes precursors in bone marrow, such as the monoblast and promonocyte, and the circulatory blood monocyte. A variety of mononuclear phagocytes is also resident in different tissues and this group of cells was formerly referred to as the reticuloendothelial system (*Figure 2.2*).

Unstimulated resident macrophages are fairly quiescent cells, but they can phagocytose and degrade particles and macromolecules via membrane receptors

18 Antigen-presenting cells

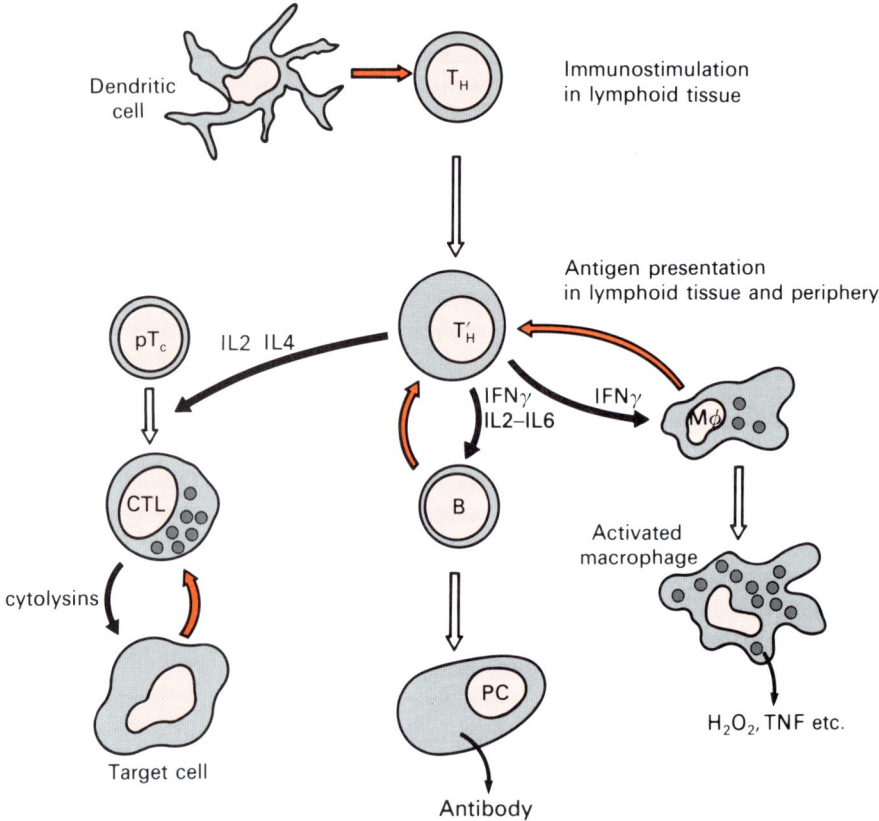

Figure 2.1. Immunostimulatory cells, notably dendritic cells, can present antigen to resting T lymphocytes and activate them (immunostimulation). APCs such as macrophages and B cells can subsequently present antigens to the activated T cells (orange arrows). Activated T_H cells in turn produce lymphokines (black arrows) which act on the APCs and other immunologically active cells, like precursor T_C cells (pT_C) which are thereby induced to mature into cytotoxic lymphocytes (CTL). The latter can secrete cytolysins which kill target cells: thus the target cells act as APCs for CTLs. TNF, tumour necrosis factor; IL, interleukin; IFN, interferon.

such as the mannosyl–fucosyl receptor, which binds some yeasts, and the Fc and complement receptors. Some resident cells constitutively secrete enzymes like lysozyme which digests bacterial cell wall peptidoglycans, but in this state they contribute relatively little to host defence. Unstimulated macrophages are susceptible to infection by a variety of intracellular parasites and, in the absence of a cell-mediated response, serious disease and death of the host may follow. The interaction of macrophages with T cells is therefore an essential part of host defence.

In the presence of an inflammatory stimulus, macrophages develop into 'stimulated' or elicited cells. Inflammatory cells have vastly increased secretory

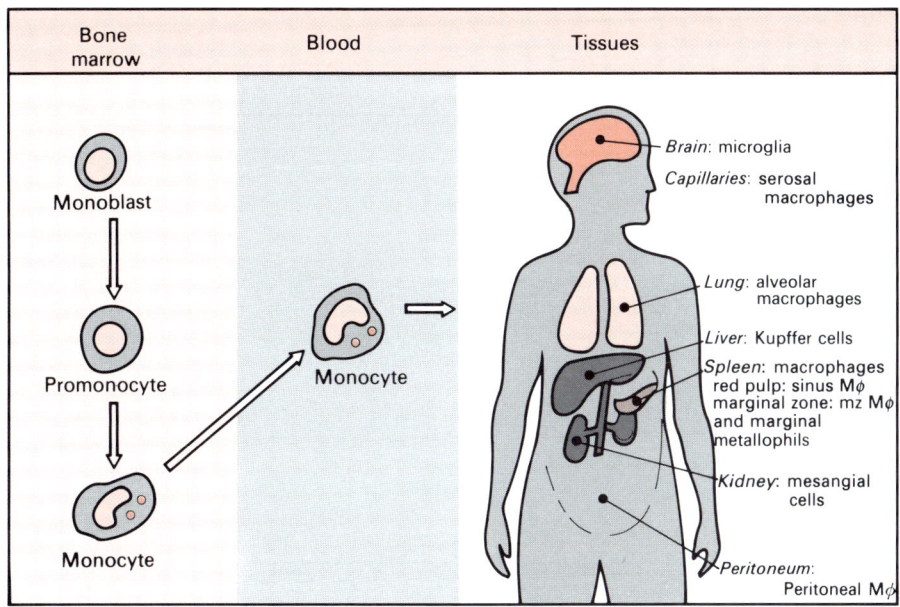

Figure 2.2. Cells of the mononuclear phagocyte lineage are derived from stem cells in the bone marrow, where they mature into monocytes. Macrophages are widely distributed throughout the body and acquire a characteristic function and phenotype depending on their location in the tissues (examples are shown). They are usually long-lived cells whose functions can be modulated depending on their environment and the presence of inflammatory stimuli or immune responses.

activities, and over 50 substances produced by them have been defined, including enzymes that digest the extracellular matrix, and molecules involved in wound healing, normal growth and tissue remodelling (1). In addition they secrete several complement components, coagulation factors, and cytokines. But even these cells have a limited capacity to deal with infectious agents.

It is only in the presence of cell-mediated immune responses that macrophages become 'activated', and although these cells are less secretory in general, they acquire cytocidal and microbicidal activities not seen in resident or inflammatory cells (*Table 2.1*). Activated cells have a number of oxygen-dependent mechanisms for killing that involve production of toxic 'reactive oxygen intermediates', particularly H_2O_2. Oxygen-independent cytotoxicity is mediated partly by tumour necrosis factor-α (TNFα) or cachectin, which has multiple effects on the body and gives rise to the 'cachexia' or wasting seen in some infections and malignancies. Activated macrophages also have an altered phenotypic profile (2). For example, expression of MHC class II molecules is greatly increased, and there are changes in the proportions of different Fc receptors on the cell surface: the mouse FcRI receptor, which binds IgG2a, is up-regulated while FcRII, which binds IgG1 and IgG2b-coated particles, is down-regulated. Moreover, the cellular consequences of ligand-binding and internalization through these receptors are different. The former mediates antibody-dependent cellular

Table 2.1. Properties of freshly isolated resident, inflammatory and activated mouse peritoneal macrophages[a]

Property	Resident	Inflammatory	Activated
Growth (CSF-dependent)	−	+	+
Spreading	−	↑	↑
Antigen F4/80	+	↓	↓
Antigen 7/4	−	−	+
MHC class II	−	+	+ +
Fc receptor	FcRII	FcRI and FcRII	FcRI
CR3	[b]+	+	+
Mannosyl–fucosyl receptor	+	+ +	↓
5′ nucleotidase	+	↓	↓
Plasminogen activator	−	+	+
Respiratory burst (PMA)			
O_2^-	−	+	+
H_2O_2	−	−	+ +
TNFα	−	−	+ +
Prostaglandin release	+ +	↓	↓
Apolipoprotein E release	+ +	↓	↓

[a]Based on ref. 32.
[b]Complement receptor 3.

cytotoxicity by activated macrophages (in part through release of H_2O_2) while the latter mediates phagocytosis and the release of arachidonic acid metabolites (e.g. prostaglandins and leukotrienes) in resident and inflammatory cells.

The ability of macrophages to present antigens may depend not only on antigen uptake, but also on the way it is processed by the cells. Biozzi mice are sometimes cited in evidence for this. These strains were inbred over more than 20 generations to produce lines which gave high or low antibody responses to a variety of antigens. Macrophages from the high responder line were found to degrade antigens less efficiently than the low responder line, so it was argued that persistence of antigens on macrophages in the former case led to more efficient presentation, although this was never really proven.

2.2 Interferon-γ in macrophage activation

The changes in macrophage phenotype and function induced during T cell responses (Section 2.1) can be mimicked by culturing resident or inflammatory cells in the presence of interferon-γ (IFNγ) (3). This lymphokine is secreted by T_H cells when they recognize foreign antigens on the macrophage, an interaction that was also shown to be MHC-restricted (4). The critical role of T cells in macrophage activation is illustrated by the wide variety of pathogenic bacteria, fungi and protozoa that can infect patients with acquired immunodeficiency syndrome (AIDS). These arise because the causative agent, human immunodeficiency virus (HIV), leads to greatly depressed helper cell function and a drastic impairment in IFNγ production. Consequently, opportunistic agents survive and replicate in the patients' macrophages although these cells are 'normal' in that they can be activated in culture by exogenous IFNγ and are then able to destroy a number of pathogens (5).

The importance of macrophage activation due to IFNγ release has been demonstrated for several bacterial infections, including *Legionella pneumophila* (6), *Mycobacterium leprae* (7) and *Listeria monocytogenes* (8). Similarly, resident macrophages are permissive for some protozoal infections, such as *Trypanosoma cruzi*, *Toxoplasma gondii* and *Leishmania*. These parasites can evade destruction in macrophages by various strategies, but they are killed if the macrophages are activated by a concomitant T cell response *in vivo* or by the addition of IFNγ *in vitro* (9).

2.3 Delayed-type hypersensitivity

Recruitment and activation of macrophages at sites of infection and inflammation is important in delayed-type hypersensitivity (DTH) reactions. These occur when sensitized individuals are re-exposed to antigen and different types of DTH may be distinguished. Contact sensitivity is characterized by swelling within 2–3 days at the site of contact with antigens such as nickel compounds. DTH can be transferred by T cells from sensitized individuals but not by antibody. Although some sensitizing agents (e.g. nickel) are too small to stimulate T cells directly they can become bound to endogenous proteins, altering their structure sufficiently to render them antigenic. The initial sensitization to antigen is probably effected by Langerhans cells in the skin (see Chapter 3), but the effector phase of the DTH response involves infiltration of the dermis and epidermis by T cells and monocytes. Sensitized T cells probably recognize APCs locally and release lymphokines that recruit and activate inflammatory cells. Cytotoxic T cells may also contribute to tissue damage in these reactions.

3. Immune rejection of allografts

DTH-like mechanisms and T_C cells may play important roles in allograft rejection (*Figure 2.3*). The importance of T cells is shown by studies on nude mice which lack a functional thymus and do not have mature T cells, although T precursors are present. They are unable to reject allografts and even xenografts (from different species) unless they receive a thymus transplant or mature T cells.

Although T cells are required for graft rejection the relative contribution of the helper and cytotoxic populations is controversial. Two basic mechanisms have been proposed although they are not mutually exclusive (10). In a 'DTH-like' process, CD4$^+$ T cells may be sensitized against MHC class II molecules on a few dendritic leukocytes from the graft (see Chapter 5), and the sensitized T cells then secrete lymphokines that recruit and activate macrophages which damage the tissue. A second mechanism involves recognition of MHC class I molecules on the graft by CD8$^+$ T_C cells and the release of cytolysins.

Different experimental approaches support these two mechanisms. In one approach T cell subsets were purified according to phenotype and transferred

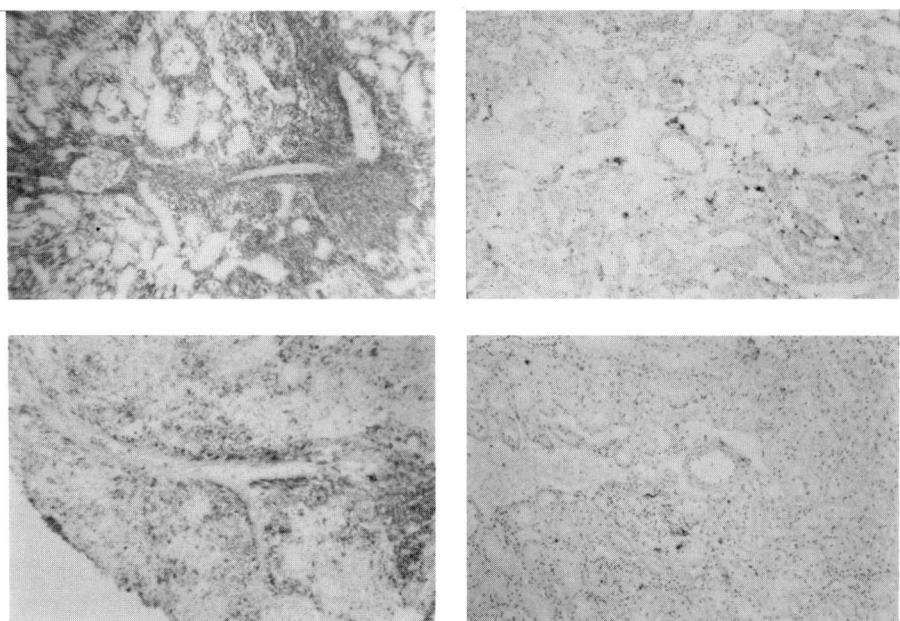

Figure 2.3. Mononuclear cells in normal and rejecting kidneys. Frozen sections are shown of rejecting rat renal allografts (left) or normal kidneys (right) stained with monoclonal antibodies against the CD45 leukocyte common antigen (top) or T cells (bottom) which were visualized by the immunoperoxidase technique (black spots). Note the massive infiltrate of leukocytes in the rejecting allografts (top left) compared to the normal kidney (top right). Many of the infiltrating cells are T lymphocytes (bottom left) which are infrequent in normal kidneys (bottom right). Sections courtesy of M.Dallman.

to adult thymectomized, irradiated, bone marrow-reconstituted rats. Such 'ATXBM' animals do not have mature T cells. Transfer of CD4$^+$CD8$^-$ cells, but not CD4$^-$CD8$^+$ cells, restored their ability to reject allografts (11). Because many T$_H$ cells are present in the CD4$^+$ subset, this was taken as evidence for a DTH-like rejection mechanism, despite the fact that class II-restricted T$_C$ cells could also be present. On the other hand, cloned T$_C$ cells specific for graft antigens triggered skin graft rejection when they were administered to mice (12). Studies with tetraparental (allophenic) mice, with mixed fur of different colours, have also shown that the effector mechanism(s) for rejection can be specific, since hair follicles of only one colour were killed when the skin was transplanted to the opposite parental strain (13). The fact that 'bystander killing' did not occur was taken as evidence for the action of T$_C$ cells rather than non-specific release of lymphokines and a generalized local reaction (i.e. DTH).

It may be that both helper and cytotoxic T cells are involved in allograft rejection. Helper T cells are often needed for CTL to develop from their precursors in culture, and this may be true *in vivo* (14). The helper requirement can be replaced by adding defined lymphokines such as IL2 and/or IL4, and perhaps other 'cytotoxic differentiation factors' (15). These lymphokines are produced when helper cells recognize APCs and may act on sensitized cytotoxic precursors to promote their development into mature CTL.

3.1 Interferon-γ in allogeneic reactions

A common finding during rejection is increased or new expression of MHC molecules on cells of the graft. For example many cells of grafted tissues do not initially express class II molecules, but these can be induced after transplantation or by treatment with various cytokines of which IFNγ is perhaps the most active. Moreover, IFNγ has been shown to synergize with TNF-α in MHC class II induction on brain endothelium and, independently, these mediators can enhance MHC class I expression. It seems likely that local release of IFNγ is responsible for the increase in Ia molecules on gut and skin epithelium during graft-versus-host responses, as well as in physiological responses. Presumably, in the latter case, this increases the ability of these tissues to display foreign antigens and to interact with specific helper cells. Interferon-γ is also known to increase the ability of endothelial cells, fibroblasts and hepatocytes to kill a number of intracellular parasites (16).

4. Antigen presentation in B cell responses

It is now clear (Chapter 4) that B cells can present antigens to T cells, as well as bind antigens via membrane Ig. Studies with nude mice indicate that T cells are required for B cell responses to many antigens. Indeed, antigens can be subdivided according to whether they are able to induce antibody responses without T cell help (T-independent; TI) or whether the B cells must interact specifically with T cells (T-dependent; TD). The degree to which the former type of response is really independent of T cells is controversial (17). TI antigens are often polymeric molecules, such as polysaccharides found on the surface of pathogens, which only elicit primary-type IgM responses. On the other hand, TD antigens are usually proteins, or hapten-carrier complexes in experimental situations, that can elicit IgM and non-IgM secondary responses. This implies that T cells control isotype switching, the process whereby a B cell expressing one class of Ig (e.g. IgM) can be induced to produce another (e.g. IgG, IgA, IgE) of the same antigen-specificity, and is an example of how T cells can alter the function of one particular type of APC.

It is not clear whether distinct populations of B cells respond to TI and TD antigens, or whether these represent B cells in different states of development or activation. Nevertheless, during TD responses, B cells present antigens to specific T cells (below) so they can receive the necessary help for activation and development into plasma cells. Claman (18), Miller and Mitchell (19), and others showed that such B–T collaboration occurred during antibody responses to antigens like heterologous erythrocytes, and studies with haptens and carriers suggested that B cells could recognize the haptenic part of the antigen while

T cells recognized carrier determinants. Thus the hapten-carrier complex was originally thought to act as a molecular bridge between the two types of lymphocyte. This idea has now been superseded in favour of the view that B cells actually present processed forms of their specific antigens to T cells.

The possibility that antigen presentation is involved in T – B collaboration was raised by experiments showing that the interaction is MHC-restricted (20), although this was not initially appreciated, and by the observation that immune response genes, now known to encode Ia molecules, are expressed in B cells. Subsequently it was shown directly that B cells can act as APCs for activated T cells *in vitro* (21). The most current idea, in essence, is that B cells use their surface Ig to take up and concentrate specific antigen, which is then processed and presented in association with MHC class II molecules to T cells (22). It has been estimated that B cells accumulate specific antigen up to 10^4 times more efficiently than other antigens.

T cell help for B cell responses can be provided by lymphokines that are secreted when the T cell recognizes antigen on the B cell, or on another APC in close vicinity. There appear to be two routes for TD antibody responses. In one the B cell responds to soluble molecules directly (23), while in the other the B cell requires a close physical contact with the T cell in a 'cognate' interaction, in which the role of soluble molecules is uncertain. Evidently, however, when a B cell presents antigen to a T cell it will be preferentially stimulated by whichever lymphokines are secreted by that T cell.

The production of monoclonal antibodies against lymphokines and recombinant molecules has helped to define the mediators involved in B cell activation. These include IFNγ, IL2, IL4, IL5 and IL6. Effects of the first two cytokines on B cells are noted in Section 5 and effects of the others follow.

Mouse IL4 (24) acts on resting B cells to up-regulate expression of Ia and the CD23 IgE receptor, and IL4-treated B cells proliferate in the presence of anti-IgM. In addition, IL4 acts on lipopolysaccharide (LPS)-treated B cells to down-regulate production of IgM and IgG3 (normally induced by LPS alone) and to stimulate production of IgG1 and IgE instead.

IL5 (25) acts on activated B cells to induce proliferation, increase Ig secretion, and expression of the IL2 receptor (which is also present on activated T cells). IL5 acts on LPS-activated B cells to stimulate production of IgA at the expense of other classes noted above. IL6 (26) increases the secretion of Ig by activated B cells.

5. Two types of helper T cell

There may be two (or more) distinct subpopulations of T_H cells. Mouse T cell clones have been derived that either secrete IL2 and IFNγ in response to antigen, or IL4 and IL5. These have been categorized as T_H1 and T_H2 cells (27). Both types produce IL3 (28), which is a lymphokine that stimulates production of many types of leukocytes from the bone marrow [multi-colony stimulating factor,

Antigen presentation in immune responses 25

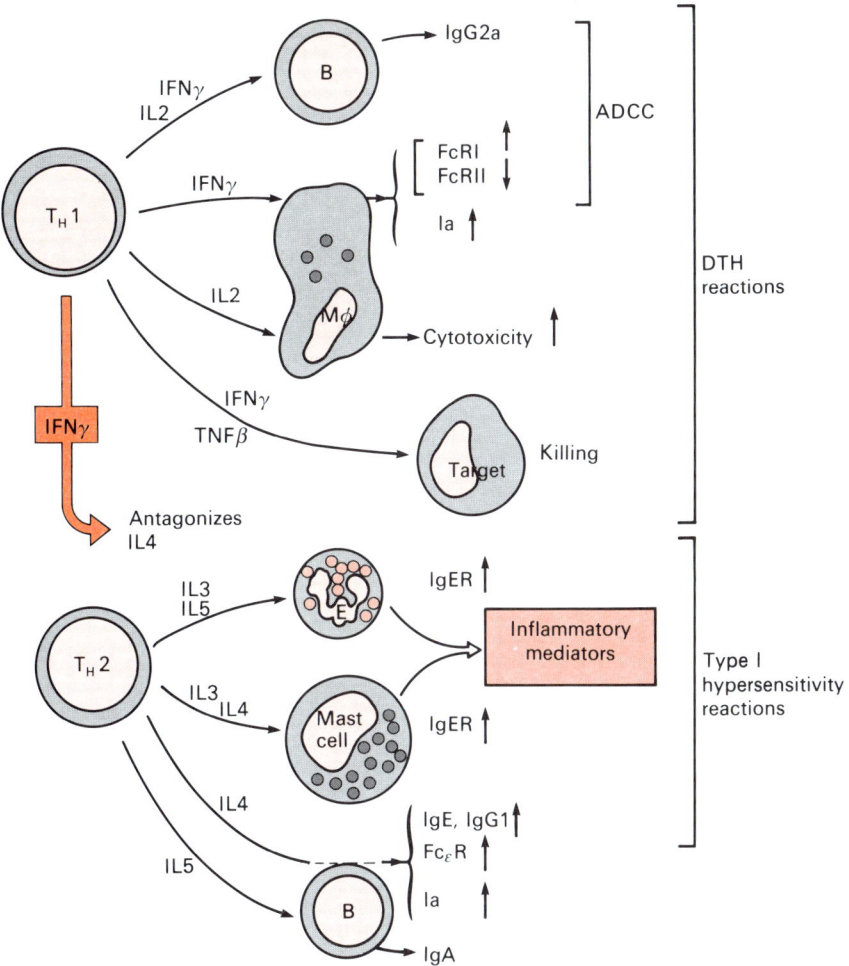

Figure 2.4. Predicted functions of T_H1 and T_H2 cells. These presumptive subsets of T cells produce different lymphokines in response to APCs, and may be important in different types of immune response, particularly DTH and type I hypersensitivity. Only some consequences of lymphokine secretion are indicated. In addition, lymphokines from the two subsets (e.g. IFNγ and IL4) may synergize, with still further effects (27). E, eosinophil; B, B cell; IgER, IgE receptor.

(CSF)], and granulocyte–monocyte CSF (GM-CSF). Further evidence that these subpopulations are present in other species has been obtained in the rat (29), where a determinant of the CD45 leukocyte common antigen, detected by the OX22 antibody, functionally discriminates T cells that make IL2 and proliferate in the mixed leukocyte reaction (MLR) (OX22$^+$) from those that do not but which provide most of the B cell help (OX22$^-$). There is evidence for similar subpopulations in humans (30).

It is not yet clear whether mouse T_H1 and T_H2 cells (like B cell subsets) represent distinct helper populations, or T cells in different states of differentiation or activation, although the latter possibility seems increasingly favoured, and it is possible they respond to different types of APC (31). Their importance is that they may contribute to different types of immune responses (*Figure 2.4*). T_H2 cells, because they produce IL4 and IL5, would be expected to stimulate IgE secretion by B cells and expression of IgE receptors. IL4 also causes proliferation of connective tissue mast cells and a combination of IL3 and IL4 stimulates bone marrow production and growth of mucosal mast cells, while IL5 induces eosinophils. Since these cells also express IgE receptors, T_H2 cells may be particularly important in immediate-type hypersensitivity reactions.

T_H1 cells produce IL2, as well as IFNγ, which antagonizes the effects of IL4 on B cells and causes IgG2a secretion. It also acts on macrophages to up-regulate expression of Ia, and of FcRI which binds IgG2a and mediates antibody-dependent cell-mediated cytotoxicity (ADCC). Interferon-γ induces IL2 receptors on some macrophages, and there is some evidence that IL2 binding increases their cytotoxicity. T_H1 cells also produce TNF-β (lymphotoxin) which can synergize with IFNγ in cytotoxic reactions. Thus, T_H1 cells may be especially important in stimulating the cidal (killing) functions of macrophages and promoting phagocytosis, and may contribute to DTH-like mechanisms. The ways in which antigens are presented to T cell subpopulations by different APCs may therefore determine the types of immunological effector mechanisms that are induced.

6. Further reading

Biochemistry of Macrophages (1986) *Ciba Found. Symp.*, **118**, Evered,D., Nugent,J. and O'Connor,M. (eds), Pittman, London.

Dallman,M. and Morris,P.J. (1988) The immunology of rejection. In *Kidney Transplantation, Principles and Practice.* Morris,P.J. (ed.), W.B.Saunders, Co., Harcourt Brace Jovanovich Inc., Philadelphia. 3rd Edn, p. 15.

De Franco,A.L. (1987) Molecular aspects of B-lymphocyte activation. *Annu. Rev. Cell Biol.*, **3**, 143.

O'Garra,A., Umland,S., De France,T. and Christiansen,J. (1988) 'B cell factors' are pleiotropic. *Immunol. Today*, **9**, 45.

Steinman,R.M. and North,R.J. (eds) (1986) *Mechanisms of Host Resistance to Infectious Agents, Tumors, and Allographs*. Rockefeller University Press, New York.

7. References

1. Takemura,R. and Werb,Z. (1984) *Am. J. Physiol.*, **246**, C1.
2. Ezekowitz,R.A.B., Austyn,J.M., Stahl,P.D. and Gordon,S. (1981) *J. Exp. Med.*, **154**, 60.
3. Nathan,C.F., Prendergast,T.J., Wiebe,M.E., Stanley,E.R., Platzer,E., Remold, H.G., Welte,K., Rubin,B.Y. and Murray,H.W. (1984) *J. Exp. Med.*, **160**, 600.
4. Rosenthal,A.S. and Shevach,E.M. (1973) *J. Exp. Med.*, **138**, 1194.

5. Murray,H.W., Gellene,R.A., Libby,D.M., Rothermel,C.D. and Rubin,B.Y. (1985) *J. Immunol.*, **135**, 2374.
6. Bhardwarj,N., Nash,T.W. and Horowitz,M.A. (1986) *J. Immunol.*, **137**, 2662.
7. Kaplan,G., Laal,S., Sheftel,G., Nusrat,A., Nath,I., Mathur,N.K., Mishra,R.S. and Cohn,Z.A. (1988) *J. Exp. Med.*, **168**, 1811.
8. Buchmeier,N.A. and Schreiber,R.D. (1985) *Proc. Natl. Acad. Sci. USA*, **82**, 7404.
9. Murray,H.W., Spitalny,G. and Nathan,C.F. (1985) *J. Immunol.*, **134**, 1619.
10. Mason,D.W. and Morris,P.J. (1984) *Annu. Rev. Immunol.*, **4**, 119.
11. Dallman,M.J., Mason,D.W. and Webb,M. (1982) *Eur. J. Immunol.*, **12**, 511.
12. LeFrancois,L. and Bevan,M.J. (1984) *J. Exp. Med.*, **159**, 57.
13. Mintz,B. and Silvers,W.K. (1970) *Transplantation*, **9**, 497.
14. Wagner,H. and Rollinghoff,M. (1978) *J. Exp. Med.*, **148**, 1523.
15. Widner,M.B. and Grabstein,K.H. (1987) *Nature*, **326**, 795.
16. Vilcek,J., Gray,P.W., Rinderknecht,E. and Sevastopoulos,C.G. (1989) In *Lymphokines*, **11**, Academic Press, New York, in press.
17. Teale,J.M. and Klinman,N.R. (1984) In *Fundamental Immunology*. Paul,W.E. (ed.), Raven Press, New York, p. 519.
18. Claman,H.N., Chaperon,E.A. and Triplett,R.F. (1969) *J. Immunol.*, **97**, 828.
19. Miller,J.F.A.P. and Mitchell,G.F. (1968) *Proc. Natl. Acad. Sci. USA*, **59**, 296.
20. Katz,D.H., Hamaoka,T. and Benacceraf,B. (1973) *J. Exp. Med.*, **137**, 1405.
21. Rock,K.L., Benacceraf,B. and Abbas,A.B. (1984) *J. Exp. Med.*, **160**, 1102.
22. Lanzavecchia,A. (1985) *Nature*, **314**, 537.
23. O'Garra,A., Umland,S., De France,T. and Christiansen,J. (1988) *Immunol. Today*, **9**, 45.
24. Noma,Y., Sideras,P., Naito,T., Bergstedt-Lindquist,S., Azuma,C., Severinson,E., Tanabe,T., Kinashi,T., Matsuda,F., Yaoita,Y. and Honjo,T. (1986) *Nature*, **319**, 640.
25. Kinashi,T., Harada,N., Severinson,E., Tanabe,T., Sideras,P., Konishi,M., Azuma,C., Tominaga,A., Bergstedt-Lindqvist,S., Takahashi,M., Matsuda,F., Yaoita,Y., Takatsu,K. and Honjo,T. (1986) *Nature*, **324**, 70.
26. Wong,G.G. and Clark,S.C. (1988) *Immunol. Today*, **9**, 137.
27. Mossman,T.R. and Coffman,R.L. (1987) *Immunol. Today*, **8**, 223.
28. Nicola,N.A. (1987) *Immunol. Today*, **8**, 134.
29. Spickett,G.P., Brandon,M.R., Mason,D.W., Williams,A.F. and Woollett,G.R. (1983) *J. Exp. Med.*, **158**, 795.
30. Rotteveel,F.T.M., Kokkelink,I., Van Lier,R.A.W., Kuenen,B., Meager,A., Miedema,F. and Lucas,C.J. (1988) *J. Exp. Med.*, **168**, 1659.
31. Hayakawa,K. and Hardy,R.R. (1988) *J. Exp. Med.*, **168**, 1825.
32. Gordon,S., Crocker,P., Lee,S.H., Morris,L. and Rabinowitz,S. (1986) In *Mechanisms of Host Resistance to Infectious Agents, Tumors, and Allografts*. Steinman,R.M. and North,R.J. (eds), Rockefeller University Press, New York, p. 128.

3

Accessory cells in culture

1. Introduction

In the 1960s it was found that antibodies were produced in culture when spleen cells, containing a mixture of B and T lymphocytes, were incubated with antigen. But a third cell type present in a fraction of glass-adherent, non-lymphoid cells was also required for these and other responses (1,2) and they were called accessory cells. Dissociated spleen cells also made antibodies when they were incubated with antigen and washed to remove the excess (i.e. 'pulsed'), despite only a small amount of antigen remaining cell-associated. Antigen-pulsed accessory cells were also effective in stimulating lymphocyte responses, but pulsed lymphocytes were not. Antigens seemed to be 'processed' to a more stimulatory form by accessory cells, and it was suggested antigens were endocytosed by macrophages (present in adherent populations) before being processed and 'presented' to specific lymphocytes. Thus, accessory cells also came to be known as antigen-presenting cells (APCs). The discovery of major histocompatibility complex (MHC) restriction led to the idea that processed antigens become associated with Ia molecules on macrophages and stimulated helper T_H cells. Since class II-restricted helper cells are often required for effector cells to develop in immune responses [e.g. activated macrophages, cytotoxic lymphocytes (CTLs) and plasma cells] APCs were generally thought of as primarily MHC class II-bearing cells.

Currently, the terms accessory cell and APC are often used interchangeably. What this does not take into consideration are more recent ideas that perhaps any cell may also become an APC for usually class I-restricted CTLs. It also tends to ignore the important distinction (Chapter 1) that activation of resting T cells (immunostimulation) is a different process to antigen presentation to sensitized T cells, and may require specialized accessory cells. These concepts are further explored in this chapter.

2. Immunostimulatory cells: dendritic cells

2.1 Features
Adherent accessory cells from lymphoid tissues (e.g. mouse spleen) were found to contain a distinct cell type in addition to the more numerous macrophages

(3). These cells were irregular in shape and actively produced and retracted cell processes, in contrast to macrophages which were more sessile and rounded (*Figure 3.1a,b,c*). Because of their origin and arborizing processes they were called lymphoid dendritic cells (DCs) and methods were developed for their purification (4). For example mouse splenic DCs only transiently adhere to tissue culture surfaces, have a low buoyant density, and lack Fc receptors, so they can be separated from macrophages which are more adherent, have a higher density, and form rosettes with antibody-opsonized particles. Using these and other techniques, DCs have been isolated from lymphoid organs of several species, and from human peripheral blood.

By a variety of criteria, DCs are members of a distinct cell lineage and are clearly different to mononuclear phagocytes. For example, DCs are non-phagocytic in culture, have a limited pinocytic capability, very few lysosomes, and lack several cytochemical markers of macrophages. DCs do not express a number of antigens present on macrophages (e.g. mouse F4/80) and granulocytes, or membrane components of T cells (e.g. T cell receptor – CD3 complex) and B cells (e.g. membrane Ig) (*Table 3.1*). They constitutively express high levels of MHC class II (and class I) molecules, the leukocyte common antigen (reflecting their bone marrow origin), and a mouse DC-restricted antigen defined by antibody 33D1. The latter was found to mediate complement-mediated cytotoxicity, which proved of immense value in functional studies on DCs.

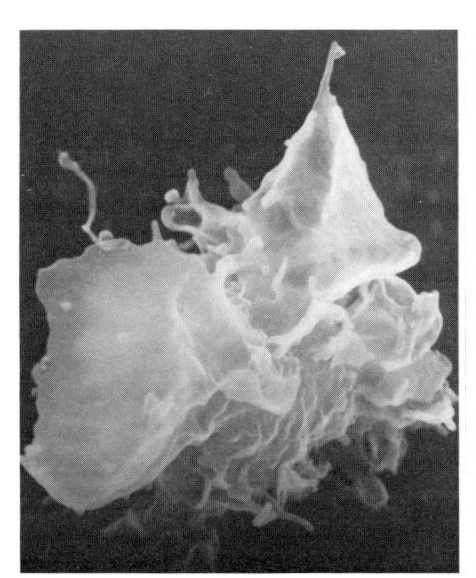

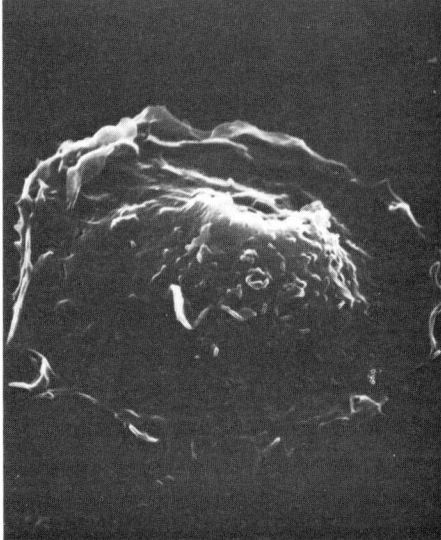

Figure 3.1a. Scanning electron micrographs of a dendritic cell in suspension (left) and a macrophage adhering to a glass surface (right). Note the irregular surface of the DCs and the presence of two flattened lamellipodiae or 'veils'. In contrast, the shape of the macrophage resembles a 'fried-egg' with circumferential membrane ruffling. Photographs courtesy of G.MacPherson and S.Gordon, respectively, Sir William Dunn School of Pathology, University of Oxford.

30 Antigen-presenting cells

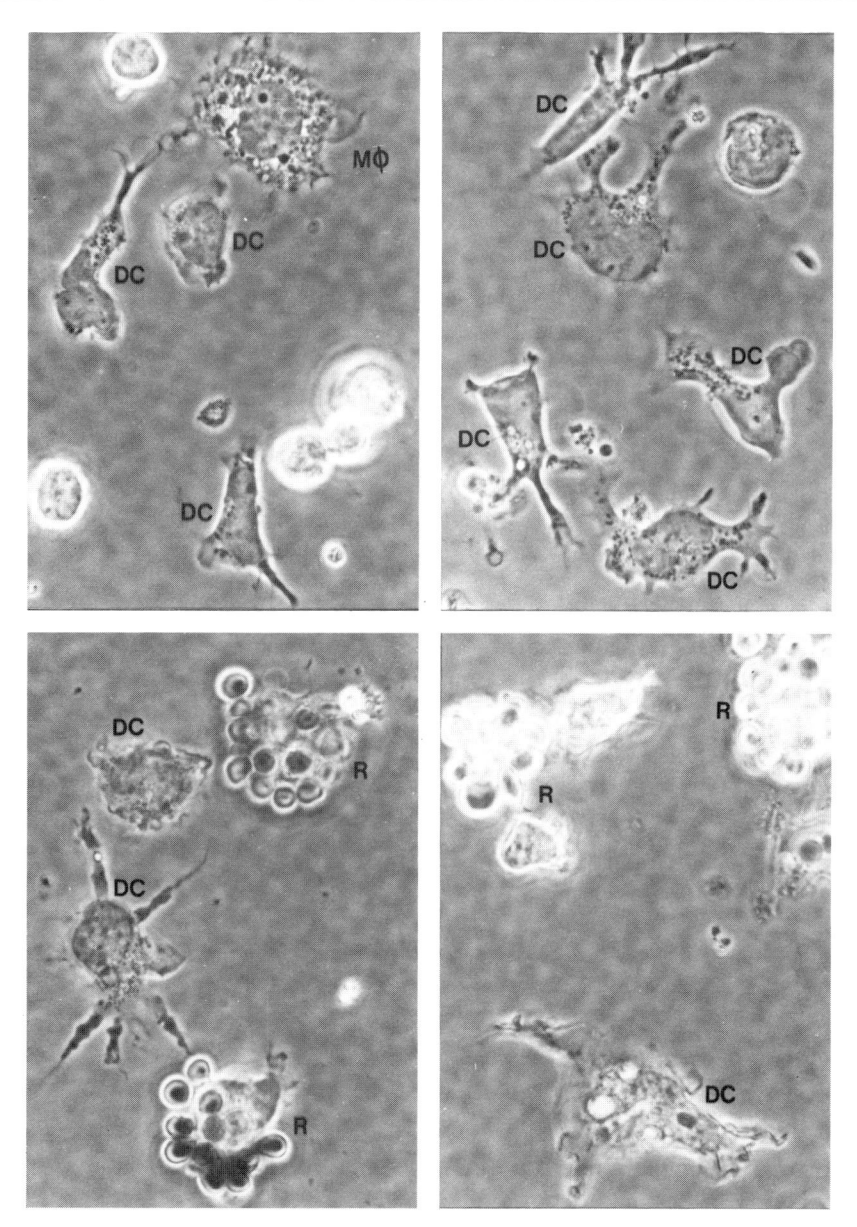

Figure 3.1b. Phase-contrast micrographs of spleen adherent cells containing dendritic cells (DCs) and macrophages before (top) and after (bottom) rosetting with antibody-coated erythrocytes. DCs are usually very irregular with a variety of dendritic processes, and contain small round phase-dense mitochondria. Macrophages (Mϕ) are more rounded and contain a number of refractile cytoplasmic inclusions and vacuoles, mostly endocytic in origin. They also have Fc receptors and form rosettes (R) with antibody-coated erythrocytes whereas DCs do not. The other cells cannot be identified with certainty here, but are probably a mixture of B cells and monocytes.

Accessory cells in culture 31

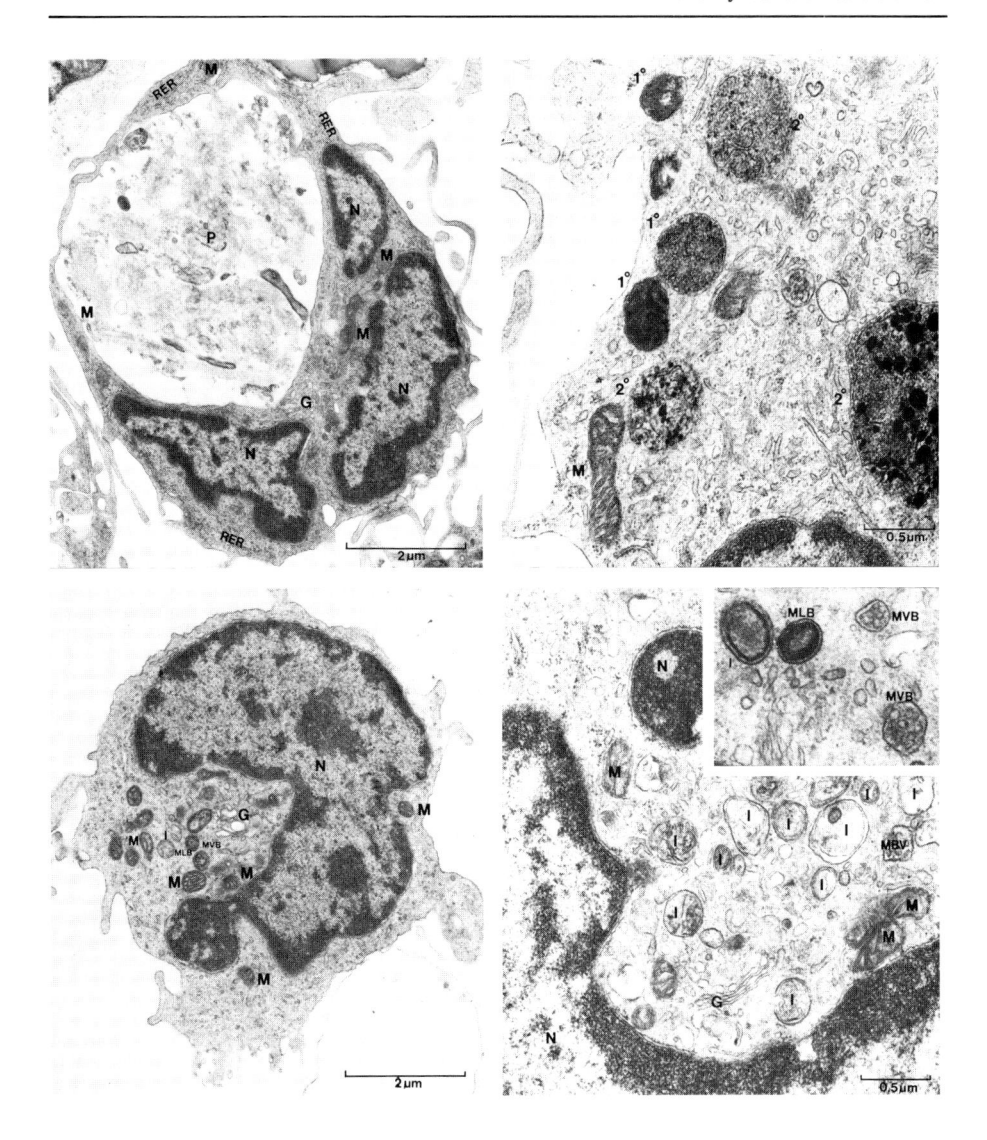

Figure 3.1c. Transmission electron micrographs of splenic dendritic cells (left) and macrophages (right). Within the cytoplasm of the dendritic cell (top left) can be seen the nucleus (N), Golgi region (G) and several round mitochondria (m), as well as multilamellar bodies (MLB), multivesicular bodies (MVB) and other inclusions (I) which are more evident in the higher-power view of another DC (bottom left). The macrophage (top right) contains a large membrane-bounded phagosome (P) and appears to be in the process of engulfing inorganic material from the preparation (extreme right); note the rough endoplasmic reticulum (RER) which is much sparser in DCs. Within the cytoplasm of another macrophage (bottom right) can be seen a spindly mitochondrion and presumptive primary (1°) and secondary (2°) lysosomes. Photographs courtesy of D.Ferguson, JRII, Oxford.

32 Antigen-presenting cells

Table 3.1. Some membrane markers of lymphoid dendritic cells

Expressed		Not detectable	
33D1 NLDC 145[a] MIDC 8[a]	Dendritic cell restricted antigens	F4/80 (mouse) CD14 (human)	Macrophage markers
MHC class I MHC class II	(Qa, Tla not tested) (including DP, DQ and DR in human)	CD1, CD3 CD4, CD5 CD8	T cell markers
LFA-1 (weak) CR3 (weak) p150,95 (human IDC)	CD18$^+$ adhesion molecules	membrane Ig CD45 (B cell type)	B cell markers
CD45 (macrophage type)		2.4G2 (mouse FcRII) CD16 (human)	Fc receptors
IL1 receptor? IL2 receptor	Lymphokine receptors		

[a]Ref. 43.

2.2 Functions

2.2.1 Mixed leukocyte reaction (MLR)

The first evidence that DCs have an important role in immune responses came from studies of the MLR in culture (5), in which alloreactive T cells (the responders) proliferate when they are cultured for a few days with allogeneic leukocytes (the stimulators, which are normally irradiated or treated with mitomycin c to prevent their proliferation).

  Enriched populations of mouse DCs were potent stimulators of the MLR, and the activity of spleen adherent cells (~60% macrophages and 30% DCs) or unfractionated spleen cells (~50% B cells and <1% DCs) correlated with DC content. When these mixed populations were treated with 33D1 and complement to kill DCs, the MLR was drastically reduced or ablated, indicating that the remaining Ia-positive leukocytes (macrophages and B cells) contributed little to the response (6) (*Table 3.2*). Macrophages cultured in interferon-γ (IFNγ) to induce high levels of Ia were also ineffective stimulators so lack of activity was not simply due to weak expression of class II MHC molecules. Similar observations were made in the human when enriched populations of monocytes, macrophages, B cells and DCs from peripheral blood or tonsil were compared (7). In the converse experiment to the mouse, treatment of peripheral blood cells with an antibody and complement to kill monocytes did not diminish accessory function and even increased it (8).

  These observations show that DCs rather than other leukocytes stimulate the allogeneic MLR. They also stimulate the syngeneic MLR (a class II-restricted proliferative response perhaps to undefined antigens that occurs when T cells are cultured with syngeneic DCs) and a large number of other responses *in vitro* (*Table 3.3*). In virtually all cases, other leukocytes such as macrophages and B cells were weak or ineffective accessory cells.

Table 3.2. Inhibition of the MLR by depleting DC. Stimulator cells were treated in the absence (−) or presence of anti-DC antibody and complement (C) and titrated into 5×10^5 allogeneic T cells in culture. The proliferative response was measured by incorporation of [^{3}H]thymidine into newly synthesized DNA (c.p.m.) Data from ref. 14, where other control groups are also shown. Stimulation is related to the number of DCs and is greatly reduced when DCs are killed by treatment with specific antibody and complement

Stimulating cells	Treatment	T cell response (c.p.m.) to stimulators ($\times 10^{-3}$)			
		24	12	6	3
Spleen adherent cells 60% Mϕ[a], 30% DCs	−	87 524	61 697	44 347	24 727
	Anti-DC + C	2 707	1 150	790	523
		320	160	80	540
Unfractionated splenocytes <1% DCs	−	60 739	30 880	12 763	2 079
	Anti-DC + C	14 874	3 064	655	355

[a]Mϕ, macrophages.

Table 3.3. Stimulation of immune responses by lymphoid dendritic cells *in vitro*

Response	Conditions
Allogeneic MLR	Alloreactive T cells proliferate when cultured with allogeneic DCs
Syngeneic MLR	Class II-restricted T cells proliferate when cultured with syngeneic DCs (autoreactivity or response to antigens in culture ?)
Development of CTLs	Alloreactive CTLs develop in the allogeneic MLR; antigen specific CTLs develop when cultured with hapten-modified syngeneic DCs
T-dependent Ig responses	Antibody-secreting cells develop when B cells are cultured with hapten/carrier complexes, T helper cells and DCs
Mitogen responses	Polyclonal proliferation occurs when oxidized T cells are cultured with DCs (oxidative mitogenesis) or when T cells are cultured with Con A or PHA and DCs
Memory responses	Memory T cells proliferate when they are restimulated with specific antigen and DCs

2.2.2 Cytotoxic lymphocytes

Dendritic cells are required for CTLs to develop in culture against class I allo-antigens in the MLR or hapten-modified syngeneic cells. CTL development often needs 'help' from class II-restricted T$_H$ cells which secrete lymphokines that act on cytotoxic precursors perhaps after antigen sensitization. However, it is possible to generate an appreciable MLR in the absence of class II differences, for example between the bm1 mutant mouse and the wild-type B6 strain which differ by three amino acids in the H-2K (MHC class I) molecule, and CTL responses can occur in the absence of CD4⁺ cells *in vivo* and *in vitro* (9,10). This implies there may be helper-independent pathways for CTL activation, although class I-restricted helper cells might contribute to the response. DCs may be directly involved because they can activate purified CD8⁺ cells (11). Thus DCs seem to be needed to trigger responses of both helper and cytotoxic T cell subsets.

2.2.3 B cells

B cells can make antibodies against T-dependent (TD) antigens by two routes, both of which require DCs. Responses against some antigens, such as epitopes on heterologous erythrocytes, can be driven by lymphokines that are secreted by helper cells when they recognize APCs other than B cells. Thus, lymphokines generated in the syngeneic MLR stimulated antibody responses when purified B cells were cultured with antigen (12). Responses to other antigens, such as hapten-carrier conjugates (e.g. trinitrophenyl–keyhole limpet haemocyanin; TNP–KLH), require physical contact between activated T blasts and histocompatible responding B cells (13). Therefore, DCs stimulate T cells to produce lymphokines for the first route, and to become lymphoblasts for cognate interactions in the second.

2.2.4 Mitogens

Accessory cell requirements for polyclonal T cell mitogenesis depend on the response in question. DCs are required for oxidative mitogenesis, in which T cells treated with oxidizing agents like sodium periodate proliferate when they are cultured with DCs but not other leukocytes (14). Responses to mitogens like concanavalin A (Con A) are sometimes stimulated better by DCs than other leukocytes, but the requirement may not be so stringent. CD3 responses (in which T cells proliferate when cultured with monoclonal antibodies to the CD3 antigen) require accessory cells with Fc receptors, which mature lymphoid DCs do not express (15,16).

2.2.5 Antigen-specific memory responses

The foregoing are examples of responses that can be elicited from unsensitized, resting T cells by DCs. DCs are also needed for restimulation of memory cells. For example, T cells from individuals immunized against tetanus toxoid or Bacillus Calmette Guerin (BCG), or infected with *Candida albicans*, proliferate in response to these antigens when DCs are added (7). Macrophages and B cells, in the mouse and human, seem unable to stimulate secondary responses *in vitro*, and macrophages are sometimes suppressive partly because they secrete prostaglandins.

2.2.6 Summary

The vast majority of data indicate that DCs are immunostimulatory cells which are required to activate resting T cells and initiate responses in culture. Other leukocytes, such as mouse splenic macrophages and resting B cells, are not stimulatory, but this is not because they express low numbers of class II MHC molecules. The most economical hypothesis is that DCs can present antigens to T cells but in addition they provide a unique activation signal(s) (Chapter 4).

3. Antigen-presenting cells

The idea that many leukocytes may be unable to initiate immune responses (Section 2) is not actually paradoxical with their possible role as APCs. This becomes clear if the accessory cell requirements of resting and sensitized T cells are considered separately.

Resting T cells are typically obtained from animals that have not been immunized against the antigen in question. Sensitized T cells, however, are often isolated after immunization, for example from draining lymph nodes or spleen following subcutaneous or intravenous administration of antigens. To a first approximation, T cell clones and hybridomas might also be considered as sensitized cells, since most clones can respond to exogenous interleukin-2 (IL2) (resting T cells do not) and hybridomas are prepared by fusing activated T cells to lymphoma lines. If this distinction is made, most data indicate that DCs are required to activate resting T cells, whereas other cell types can present antigens to sensitized T cells. Thus macrophages and resting B cells are not immunostimulatory but can present antigens to sensitized helper cells, as can fibroblasts transfected with the appropriate MHC class II molecules (which are not normally expressed) (17).

The essentially 'passive' nature of antigen presentation is evident from the fact that purified cell membranes, supported planar membranes, and liposomes are all able to 'present antigens' to sensitized T cells provided they contain the requisite MHC molecules (18,19). Immunostimulation, however, requires viable DCs since heat-treated or UV-irradiated cells are inactive. Moreover, the response of a sensitized T cell to APCs is proportional to the concentration of antigen and density of MHC molecules (20), which is clearly not the case for immunostimulation since macrophages can express even more class II molecules than DCs, but do not function.

The respective roles of DCs and other APCs in T cell responses is evident from their differing abilities to stimulate resting T cells and (CD8$^-$) alloreactive blasts produced in the MLR (21). Resting T cells respond only to allogeneic DCs, but sensitized T cells can be stimulated by allogeneic macrophages, resting B cells, and DCs (*Figure 3.2*).

Resting B cells can act as APCs but are not stimulator cells in the MLR (above). However, some studies have shown that activated B blasts and the A20 B cell line can stimulate the MLR (22). Therefore, other leukocytes may have immunostimulatory activity as well as DCs, although it is hard to understand how B blasts could contribute to the induction of a physiological response. Conceivably, B cells are first activated by T_H cells and DCs in one area, and

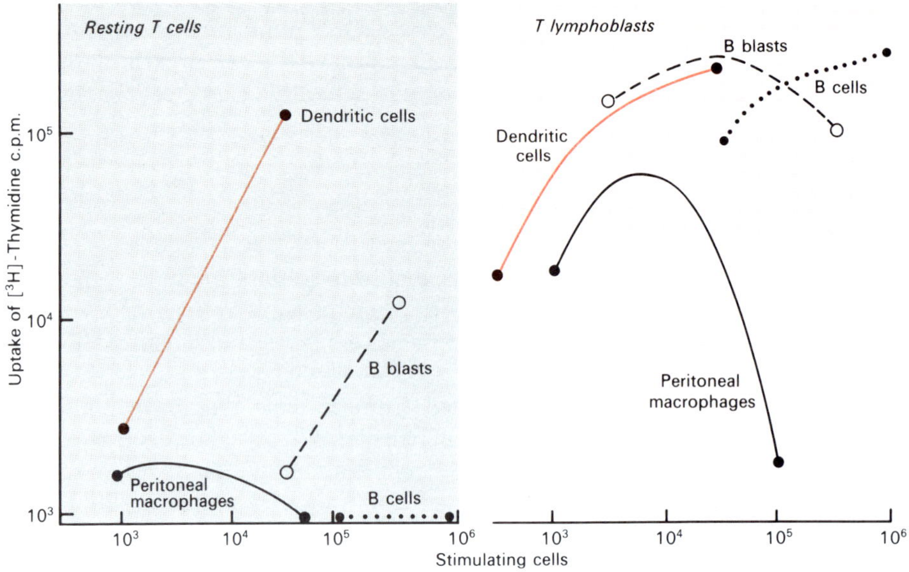

Figure 3.2. Populations of dendritic cells, peritoneal macrophages, B cell blasts and B cells were tested for their ability to stimulate proliferation ([^{3}H]thymidine incorporation) of resting T cells in the MLR (left) and alloreactive CD8$^-$ lymphoblasts from the MLR (right). The macrophages had been cultured in IFNγ to induce high levels of Ia. Note macrophages do not stimulate resting T cells, but are accessory cells for lymphoblasts, although they are inhibitory at high doses. The weak activity of B cell blasts may be real or due to contaminating DCs. (Data from ref. 21.)

B blasts then activate specific resting T cells in another.

Some non-leukocytes, such as epithelial and endothelial cells, are induced to express Ia molecules during immune responses *in vivo*, or by exogenous IFNγ *in vitro*. These cells should be able to present antigens to (class II-restricted) sensitized T cells, but we are not aware of convincing reports that they can activate resting T cells. Even though endothelial cells, for example, reportedly stimulate the primary MLR, many preparations are known to contain a small subset of Ia-rich leukocytes, possibly DCs (23). Other cells, such as keratinocytes and fibroblasts, have been shown to be inactive in the MLR, even after Ia expression was increased by IFNγ (24,25).

4. Dendritic leukocytes *in situ*

This chapter is concluded with an examination of the distribution of dendritic cells in the body (*Table 3.4*). Collectively these cells and their precursors can be referred to as dendritic leukocytes (dendrocytes) because of their characteristic cytology and bone marrow origin.

Table 3.4. Distribution of dendritic leukocytes

Type	Localization
Lymphoid dendritic cell (DC)	(Cells isolated from lymphoid tissues and studied *in vitro*)
Interdigitating cell (IDC)	T cell areas of secondary lymphoid tissues, and thymus medulla
Follicular dendritic cell (FDC)	B cell areas of secondary lymphoid tissues
Veiled cells (VC)	Afferent lymphatics, MNLX TDL (see text)
Langerhans cells (LC)	Epidermis of skin
Non-lymphoid or interstitial dendritic leukocytes	Interstitial connective tissue of most non-lymphoid organs

4.1 Interdigitating cells of T areas

Lymphoid tissues like the spleen and lymph node are highly organized structures. Within them, T and B cells tend to be localized in discrete areas and associated with distinct types of non-lymphoid cells (*Figure 3.3*). In T areas there are interdigitating cells (IDCs), while in B areas there are follicular dendritic cells. Of these, and of all dendritic leukocytes in the body, IDCs are phenotypically the most similar to isolated lymphoid DCs (26). By immunocytochemistry, IDCs are strongly Ia$^+$ cells that do not express a variety of T cell, B cell or macrophage markers, and closely resemble lymphoid DCs, except the antigen 33D1 is undetectable. If isolated lymphoid DCs are re-injected into animals, they migrate specifically into T areas, suggesting at least a close relationship with IDCs (see Chapter 4), but under certain circumstances IDCs may be phagocytic *in situ* (27). It is possible that IDCs *per se* are not obtained during conventional isolation techniques and lymphoid DCs may actually correspond to a migratory population normally present in the marginal zone, and even red pulp, which only subsequently migrates into the white pulp to become IDCs.

IDCs are present in all lymphoid tissues, including the thymus where they are localized to the medulla, although they reportedly may also be present at the cortico-medullary junction. Evidence has been obtained for an Ia$^-$ precursor to DCs in the thymus that matures into an immunostimulatory Ia$^+$ DC, perhaps under the influence of IL1 (28).

4.2 Follicular dendritic cells of B areas

The phenotype of follicular DCs (FDCs) in B areas is much less certain, as is their origin for they may not be bone marrow-derived (although this is controversial). It has proven difficult to isolate FDCs for *in vitro* studies, and even though some reports claim success it is possible these preparations also contain lymphoid DCs.

What does seem clear is that FDCs can retain immune complexes on their surface for considerable lengths of time, even after the immune response to

38 Antigen-presenting cells

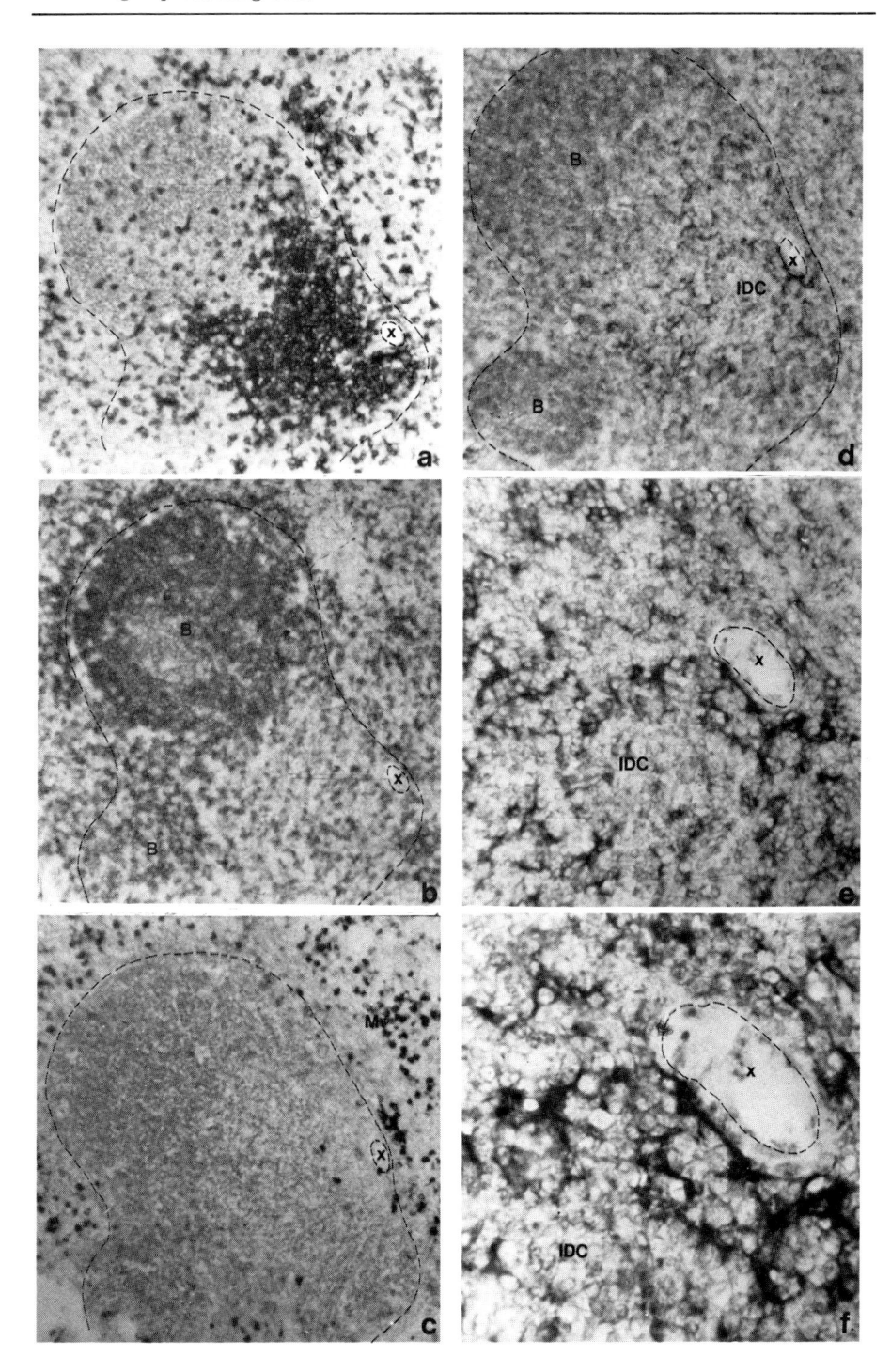

antigen in the complexes has waned (29). For example, horseradish peroxidase (HRP)–anti-HRP complexes can be detected on FDCs after intravenous injection, attachment being mediated via complement receptors: the complexes are not seen after treatment of mice with cobra venom factor which depletes the animal of complement C3. Such complexes are not detectable on IDCs.

It is interesting to speculate that IDCs express processed antigens bound to MHC molecules to stimulate T cells, whereas native antigen in complexes on FDCs may be important for some B cell responses. Antigens on FDCs may be more important for inducing B cell memory than for triggering primary responses, since immune complexes can only form after antibody has been produced (30) (unless 'natural antibodies' play a role).

4.3 Veiled cells of afferent lymph

Cells resembling lymphoid DCs but called 'veiled cells' (VCs) have been collected from afferent lymph (which drains from peripheral tissues to lymph nodes) in several species. These cells are almost undetectable in efferent or central lymph (e.g. in the thoracic duct) so VCs migrate into the node but do not emerge from it. In the rat, a technique has been developed to obtain larger numbers of these cells in greater purity. If the mesenteric nodes are surgically removed, and the lymphatics are allowed to rejoin, cells that migrate from the gut in afferent lymph appear in the thoracic duct, from where they can be obtained after cannulation. Irradiation of the animal leads to a considerable reduction in the number of lymphocytes in the lymph and a higher proportion of VCs. Such cells, from 'mesenteric lymphadenectomized, X-irradiated, thoracic duct lymph' (MNLX TDL), stimulate the allogeneic MLR and oxidative mitogenesis (31), and home to T areas of popliteal lymph nodes after injection into the footpad (32).

Veiled cells from MNLX TDL are quite heterogeneous. They have a variety of cytoplasmic projections including lamellipodiae or 'veils' from where they get their name (*Figure 3.4*), express MHC class II molecules, and are non-phagocytic in culture, although the presence of Feulgen-positive (i.e. DNA-containing) and other inclusions suggests they were phagocytic at some time in the past. However, in the rat VCs variably express CD4 (which is also found on T cells and some macrophages in rats and humans, but not in mice) and IL2 receptors

Figure 3.3. Organization of spleen. Sequential frozen sections of mouse spleen were labelled with monoclonal antibodies specific for T cells (**a**), B cells (**b**), complement receptor type 3 on macrophages (Mϕ; **c**), or MHC class II (**d,e,f**), and visualized by the peroxidase technique (the darker regions, which differ in intensity between sections). Photographs a–d are similar areas at the same magnification, while d–f are increasing magnifications of the area near the central arteriole (X) to the right of each photograph. The approximate area of white pulp is delineated by dashed lines in a–d. Note the complementary staining patterns for T and B areas (a and b) which together comprise the white pulp; macrophages are present in the red pulp (c). The irregular interdigitating cells (IDC) scattered throughout T areas are labelled with anti-Ia (d,e,f) as are small round B cells (d, compare with b; top right of e). Follicular dendritic cells in B areas are not visualized with these antibodies.

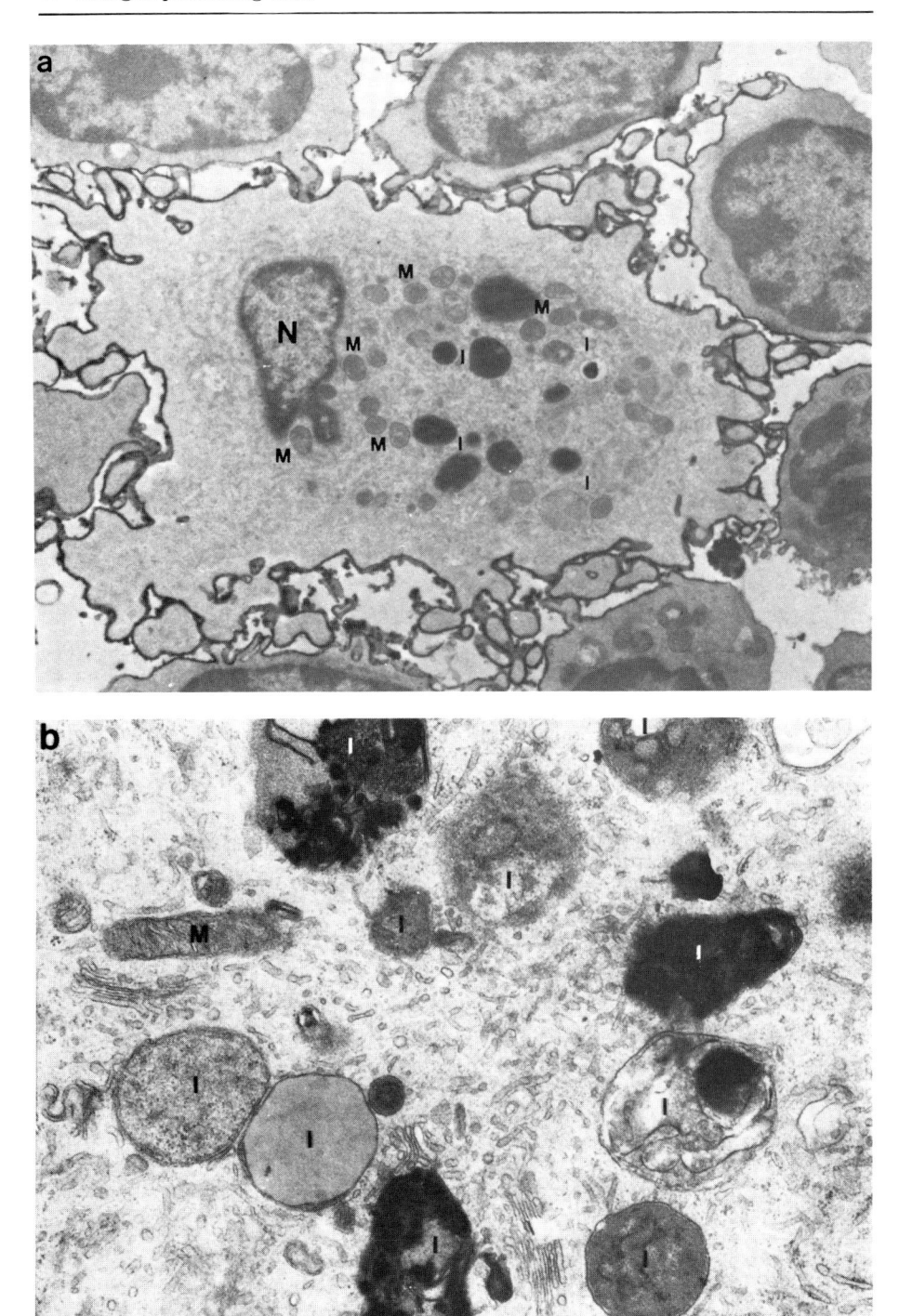

40 Antigen-presenting cells

[which can be up-regulated by culture in granulocyte – macrophage colony stimulating factor (GM-CSF)]. A subpopulation stains for ATPase and non-specific esterase which are not present in lymphoid DCs. The phenotype of VCs seems to be transitional between mature lymphoid DCs and a precursor form, the Langerhans cells of skin.

4.4 Langerhans cells of skin

Langerhans cells (LCs) are irregularly shaped, Ia$^+$ leukocytes in the epidermis of skin. They have characteristic 'racket shaped' granules in their cytoplasm called Birbeck granules which are of unknown function but may be associated with endocytic activity (33) (*Figure 3.5*). Langerhans cells can be isolated from epidermal sheets of skin by enzyme digestion and separation from keratinocytes.

Unlike lymphoid DCs, freshly isolated LCs have a high buoyant density and express Fc receptors and a few macrophage markers such as F4/80 in the mouse, as well as membrane ATPase and non-specific esterase. These characteristics disappear when the cells are cultured over several days, particularly in the presence of GM-CSF. The expression of Ia increases somewhat, but other markers like low levels of CR3 persist relatively unchanged. Cultured LCs come to resemble lymphoid DCs but they do not acquire the 33D1 antigen (34,35) (*Table 3.5*).

This phenotypic transformation is accompanied by marked functional changes. Freshly isolated LCs have very weak immunostimulatory activity, for example they are poor stimulators of the allogeneic MLR, but this increases after culture so that on a per cell basis cultured LCs are more active than splenic DCs (36). In addition, (immature) LCs are more endocytic than lymphoid DCs, and they have been reported to take up HRP *in situ* (37), and bacteria and lectins *in vitro* (33,38), as well as to internalize monoclonal antibodies to CD1 (T6, expressed on LCs and some thymocytes) and human HLA-DR (39). Langerhans cells may leave the skin as VCs in cutaneous lymphatics, and migrate to the draining nodes to become IDCs (see Chapter 5).

4.5 Dendritic leukocytes in non-lymphoid organs

Dendritic leukocytes may be present in many other non-lymphoid tissues in addition to skin epithelium. For example, irregular cells that express Ia and the leukocyte common antigen were described in the interstitium of rat heart and other solid organs, but these cells did not stain for membrane Ig (B cells) or W3/13

Figure 3.4. Transmission electron micrographs of veiled cells from rat MNLX TDL. Low power view of a veiled cell stained with anti-Ia antibody using the immunoperoxidase technique (**a**), and a higher power view of the cytoplasm of another veiled cell (**b**). Note the circumferential staining of the veiled cell (a) which contains several round mitochondria and dense inclusions and a small glancing section of the nucleus; there is artefactual staining of adjacent cells in regions of closest proximity to the veiled cell. A variety of cytoplasmic inclusions is evident in the higher power view (b). Labels as in *Figure 3.1c*.

42 Antigen-presenting cells

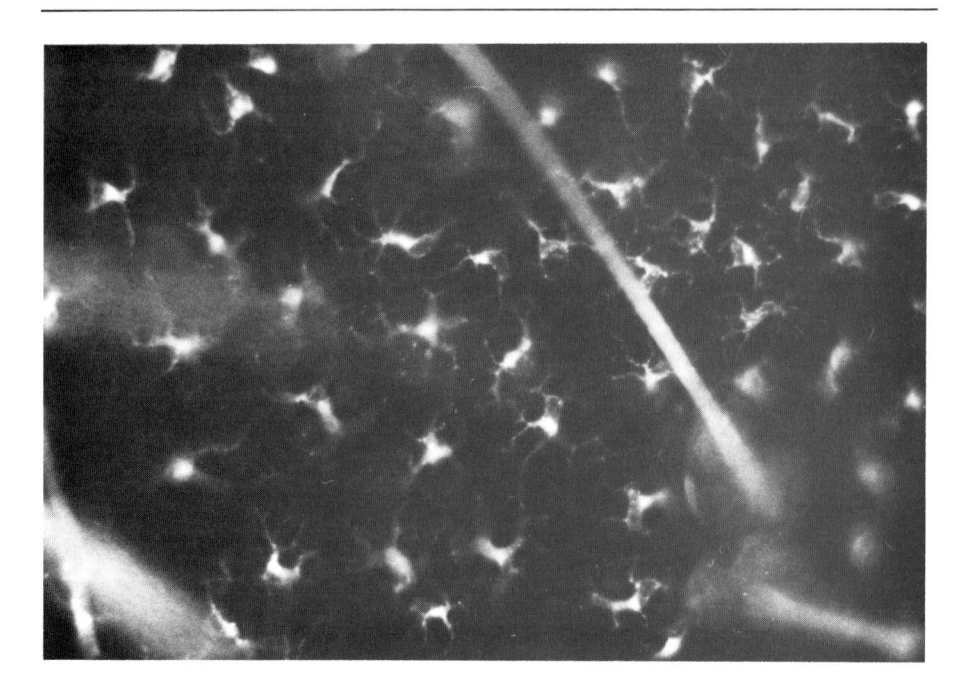

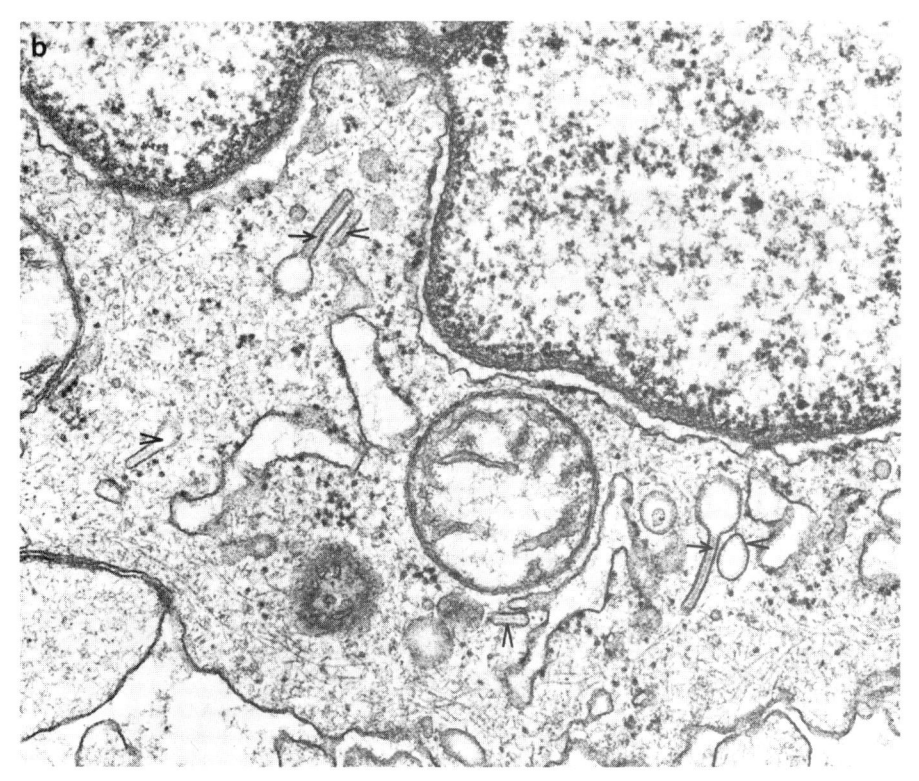

(T cells and macrophages), lacked a number of cytochemical markers of macrophages, and did not appear to be phagocytic (40). The unknown is whether these cells more closely resemble immunostimulatory lymphoid DCs or whether they are precursors. This question has not been systematically studied. Although dendritic leukocytes with stimulatory activity in the MLR have been isolated from some tissues such as rat liver and the airways of a number of species (41,42) it is possible they may have developed and acquired immunostimulatory activity during the assay (see also Chapter 5).

4.6 Summary

Dendritic leukocytes constitute a distinct lineage that may be as widely distributed throughout the body, and varied in phenotype and function, as mononuclear phagocytes. Precursors are certainly present in skin, and perhaps other peripheral tissues, and maturation can be promoted by IL1 (in the thymus) and/or GM-CSF (in the skin). Mature immunostimulatory DCs may be confined to lymphoid tissues, especially in T areas consistent with their function of activating resting T cells.

5. Further reading

Austyn,J.M. (1987) Lymphoid dendritic cells. *Immunology*, **62**, 161.
Steinman,R.M., Van Voorhis,W.C. and Spalding,D.M. (1986) Dendritic cells. In *Handbook of Experimental Immunology*. Weir,D.M., Herzenberg,L.A., Blackwell,C. and Herzenberg,L.A. (eds), Blackwell, Oxford, 4th edn, Chapter 49.
Wolff,K. and Stingl,G. (1983) The Langerhans cell. *J. Invest. Dermatol.*, **80**, 17s.
Thorbecke,G.J., Silberberg-Sinakin,I. and Flotte,T.J. (1980) Langerhans cells as macrophages in skin and lymphoid organs. *J. Invest. Dermatol.*, **75**, 32.

Table 3.5. Phenotypes of freshly isolated and cultured Langerhans cells, peritoneal macrophages and splenic DCs from mouse

Parameter	Macrophage	Fresh LCs	Cultured LCs	DCs
Birbeck granules	−	+	−	−
Non-specific esterase	+	+	−	−
Membrane ATPase	+	+	−	−
Antigen F4/80	+	+	−	−
Fc receptor (2.4G2)	+	+	−	−
Complement receptor CR3	+	+	+	+
MHC class II	+/−	+	+ +	+
Antigen 33D1	−	−	−	+

Figure 3.5. Langerhans cells in epidermis of skin. (**a**) Langerhans cells visualized within an epidermal sheet by immunofluorescence with an anti-Ia monoclonal antibody. (**b**) Transmission electron micrograph of part of the cytoplasm and nucleus of a Langerhans cell illustrating cytoplasmic Birbeck granules which have been completely (arrows) or partially (arrowheads) sectioned. Photographs courtesy of (a) C.Larsen and R.Steinman and (b) N.Romani and G.Schuler.

6. References

1. Mosier,D.E. (1967) *Science,* **158**, 1573.
2. Hersh,E.M. and Harris,J.E. (1968) *J. Immunol.,* **100**, 1184.
3. Steinman,R.M. and Cohn,Z.A. (1973) *J. Exp. Med.,* **137**, 1142.
4. Steinman,R.M., Van Voorhis,W.C. and Spalding,D.M. (1986) In *Handbook of Experimental Immunology.* Weir,D.M., Herzenberg,L.A., Blackwell,C. and Herzenberg,L.A. (eds), Oxford, 4th edn, Chapter 49.
5. Steinman,R.M. and Witmer,M.D. (1978) *Proc. Natl. Acad. Sci. USA,* **75**, 5132.
6. Steinman,R.M., Gutchinov,B., Witmer,M.D. and Nussenzweig,M.C. (1983) *J. Exp. Med.,* **157**, 613.
7. Van Voorhis,W.C., Valinsky,J., Hoffman,E., Luban,J., Hair,L.S. and Steinman,R.M. (1983) *J. Exp. Med.,* **158**, 174.
8. Van Voorhis,W.C., Steinman,R.M., Hair,L.S., Luban,J., Witmer,M.D., Koide,S. and Cohn,Z.A. (1983) *J. Exp. Med.,* **158**, 126.
9. Sprent,J. and Schaefer,M. (1985) *J. Exp. Med.,* **162**, 2068.
10. Sprent,J., Schaefer,M., Lo,D. and Korngold,R. (1986) *J. Exp. Med.,* **163**, 998.
11. Inaba,K., Young,J.W. and Steinman,R.M. (1987) *J. Exp. Med.,* **166**, 182.
12. Inaba,K., Granelli-Piperno,A. and Steinman,R.M. (1983) *J. Exp. Med.,* **158**, 2040.
13. Inaba,K., Witmer,M.D. and Steinman,R.M. (1984) *J. Exp. Med.,* **160**, 858.
14. Austyn,J.M., Steinman,R.M., Weinstein,D.E., Granelli-Piperno,A. and Palladino,M.A. (1983) *J. Exp. Med.,* **157**, 1101.
15. Smith,K.G.C., Austyn,J.M., Hariri,G., Beverley,P.C.L. and Morris,P.J. (1986) *Eur. J. Immunol.,* **16**, 478.
16. Austyn,J.M., Smith,K.G.C. and Morris,P.J. (1987) *Eur. J. Immunol.,* **17**, 1329.
17. Malissen,B., Peele-Price,M., Goverman,J.M., McMillan,M., White,J., Kappler,J., Marrack,P., Pierres,A., Pierres,M. and Hood,L. (1984) *Cell,* **36**, 319.
18. Walden,P., Nagy,Z.A. and Klein,J. (1985) *Nature,* **315**, 327.
19. Watts,T.H. and McConnell,H.M. (1988) In *Processing and Presentation of Antigens.* Pernis,B., Silverstein,S.C. and Vogel,H.J. (eds), Academic Press, San Diego, p. 143.
20. Matis,L.A., Glimcher,L.H., Paul,W.E. and Schwartz,R.H. (1983) *Proc. Natl. Acad. Sci. USA,* **80**, 6019.
21. Inaba,K. and Steinman,R.M. (1984) *J. Exp. Med.,* **160**, 1717.
22. Krieger,J.I., Chesnut,R.W. and Grey,H.M. (1986) *J. Immunol.,* **137**, 3117.
23. Nunez,G., Ball,E.J. and Stastny,P. (1983) *J. Immunol.,* **131**, 666.
24. Pober,J.S., Gimbrone,M.A., Cotran,R.S., Reiss,C.S., Burakoff,S.J., Fiers,W. and Ault,K.A. (1983) *J. Exp. Med.,* **157**, 1339.
25. Breathnach,S.M., Shimada,S., Kovac,Z. and Katz,S.I. (1986) *J. Invest. Dermatol.,* **86**, 226.
26. Witmer,M.D. and Steinman,R.M. (1984) *Am. J. Anat.,* **170**, 465.
27. Fossum,S. and Rolstad,B. (1986) *Eur. J. Immunol.,* **16**, 440.
28. Inaba,K., Witmer-Pack,M.D., Inaba,M., Muramatsu,S. and Steinman,R.M. (1988) *J. Exp. Med.,* **167**, 149.
29. Tew,J.G., Mandel,T.E., Phipps,R.P. and Szakal,A.K. (1984) *Am. J. Anat.,* **170**, 407.
30. Tew,J.G., Phipps,R.P. and Mandel,T.E. (1980) *Immunol. Rev.,* **53**, 175.
31. Pugh,C.W., MacPherson,G.G. and Steer,H.W. (1983) *J. Exp. Med.,* **157**, 1758.
32. Fossum,S. (1988) *Scand. J. Immunol.,* **27**, 97.
33. Takigawa,M., Iwatsuki,K., Yamada,M., Okamoto,H. and Imamura,S. (1985) *J. Invest. Dermatol.,* **85**, 12.
34. Schuler,G. and Steinman,R.M. (1985) *J. Exp. Med.,* **161**, 526.
35. Heufler,C., Koch,F. and Schuler,G. (1988) *J. Exp. Med.,* **167**, 700.
36. Inaba,K., Schuler,G., Witmer,M.D., Valinsky,J., Atassi,B. and Steinman,R.M. (1986) *J. Exp. Med.,* **164**, 605.

37. Wolff,K. and Schreiner,E. (1970) *J. Invest. Dermatol.*, **54**, 37.
38. Schuler,G., Aubock,J. and Linert,J. (1983) *J. Immunol.*, **130**, 2008.
39. Hanau,D., Fabre,M., Schmitt,D.A., Garaud,J.-C., Pauly,G., Tongio,M.-M., Mayer,S. and Cazenave,J.-P. (1987) *Proc. Natl. Acad. Sci. USA,* **84**, 2901.
40. Hart,D.N.J. and Fabre,J.W. (1981) *J. Exp. Med.*, **154**, 347.
41. Sertl,K., Takemura,T., Tschachler,E., Ferrans,V.J., Kaliner,M.A. and Shevach,E.M. (1986) *J. Exp. Med.,* **163**, 436.
42. Holt,P.G., Schon-Hegrad,M.A. and Oliver,J. (1987) *J. Exp. Med.,* **167**, 262.
43. Breel,M., Mebius,R.E. and Kraal,G. (1987) *Eur. J. Immunol.,* **17**, 1555.

4

Antigen processing and immunostimulation

1. Introduction

Accessory cells have three main functions in T cell responses.

(i) They process antigens to forms that can bind to major histocompatibility complex (MHC) molecules and can be recognized by T cell receptors (TCRs); these antigens may be exogenous (i.e. from the cell's environment) or endogenous (i.e. synthesized within the cell).
(ii) They express the requisite MHC molecules for T cell recognition.
(iii) If they can activate resting T cells, they presumably elaborate T cell activation signals which could be mediated by soluble molecules and/or direct membrane molecular interactions.

Accessory cells can thus be divided into antigen-presenting cells (APCs), which perform the first two functions, and immunostimulatory cells which carry out the third function (*Table 4.1*). This chapter considers how antigens are processed and presented to T cells, and discusses possible mechanisms of immunostimulation.

2. Antigen processing and presentation

Antigen processing may be related to mechanisms for the normal turnover of cellular molecules. Cell proteins, for instance, can be degraded by two main routes. Lysosomal proteolysis can occur when cells are deprived of their essential nutrients as well as in professional phagocytes, while non-lysosomal, energy-dependent proteolytic systems, including the ubiquitin pathway, are active in normally growing cells. With this in mind, this section focuses on the presentation of exogenous antigens to helper T (T_H) cells by macrophages and B cells, and

Table 4.1. Functions of accessory cells

	MHC expression	Antigen processing	Signals to activate resting T cells
Antigen presentation	+	+	
Immunostimulation	+	+	+

on the presentation of endogenous antigens to cytotoxic lymphocytes (CTLs) by fibroblasts and perhaps any cell in the body.

2.1 Macrophages

2.1.1 Endocytosis

Macrophages phagocytose particles from their environment after attachment to specific membrane receptors and by less well understood pathways (e.g. for uptake of latex beads). They also internalize soluble molecules from the extracellular fluid by pinocytosis. This can be constitutive (essentially, the random sampling of fluids) or 'regulated', meaning that uptake occurs via specific membrane receptors which enable the cell to internalize nutrients, growth factors and other molecules. Together, phagocytosis and pinocytosis are known as endocytosis.

During endocytosis, large areas of plasma membrane and considerable volumes of fluid are internalized (e.g. the macrophage internalizes the equivalent of its entire surface area every 30 min) but the area and volume of the cell remain relatively constant. This is achieved by membrane recycling: as fast as membrane-bound vesicles are internalized, there is a compensatory flow of vesicles containing fluid and solutes back to the surface. Endocytosis in macrophages is of particular interest in relation to antigen processing and presentation, because particles and molecules are often degraded after endocytosis, and membrane recycling provides a means of re-expressing internalized molecules at the cell surface.

In many cases receptor-mediated endocytosis occurs at specialized regions of the membrane (called 'coated pits' because the membrane is indented in these areas and coated on its cytoplasmic face with a structure composed partly of clathrin). Receptor–ligand complexes are often internalized as 'coated vesicles', the coat dissociates, and the resultant vesicle fuses with others to form a larger 'endosome'. Ultimately, the latter may fuse to other vesicles, including primary lysosomes which contain various proteases in an acidic environment, to form secondary lysosomes (*Figure 4.1*).

Ligands can be degraded in secondary lysosomes as well as their receptors (e.g. epidermal growth factor and insulin), or their receptors may escape degradation and be recycled back to the plasma membrane (e.g. mannose-6-phosphate receptor). Not all ligands are completely digested, and some are subject to limited modification and then recycled; in this way, for example, a proenzyme can be converted to the active enzyme and transported back to the surface. Other ligands are not modified at all, but are recycled intact. Thus, the transferrin receptor binds and internalizes iron-saturated transferrin, the iron is released in a low pH compartment, and the transferrin is transported back to the surface and liberated. The precise form of the ligand may determine the outcome: IgG monomers bound to the Fc receptor are recycled intact, whereas IgG multimers and the receptor are degraded intracellularly. Some ligands such as mannosylated proteins are not transported to lysosomes because they are modified in pre-lysosomal compartments, one of which may be the 'compartment

48 Antigen-presenting cells

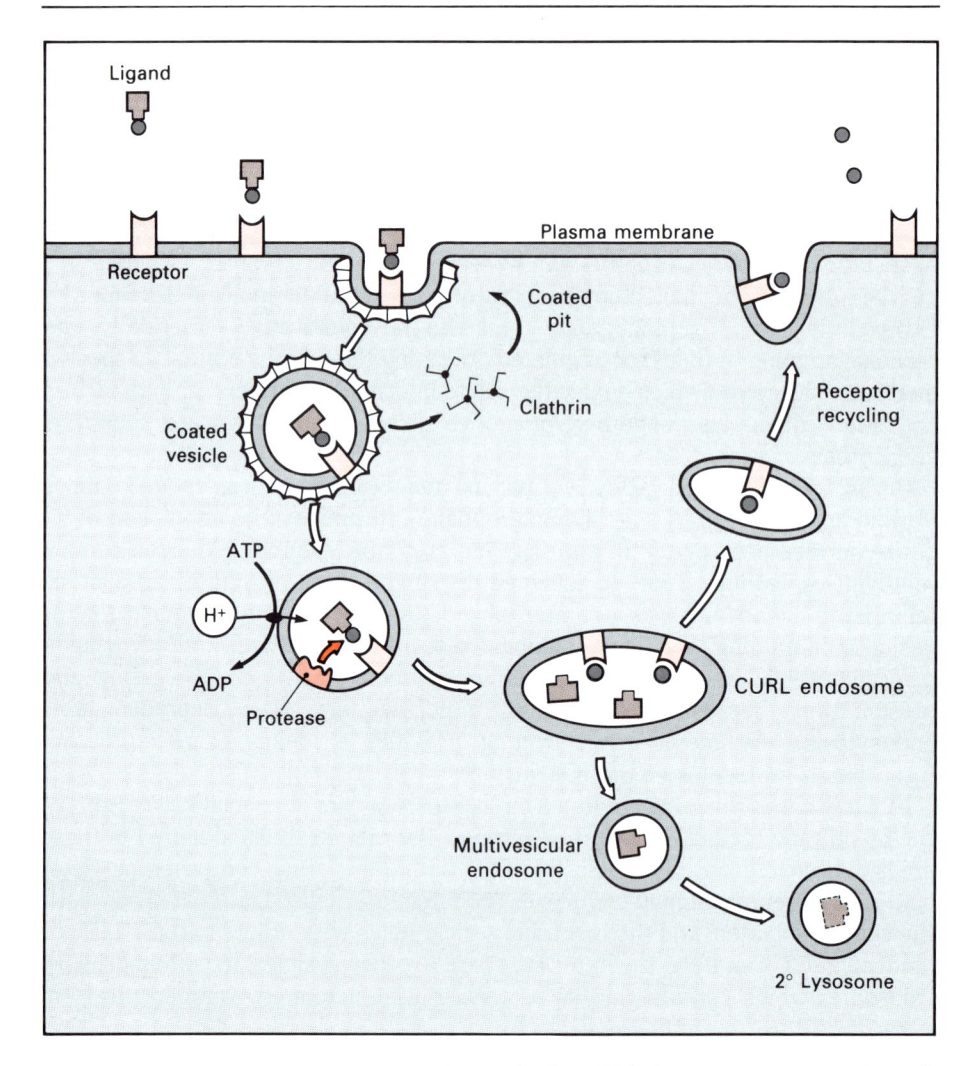

Figure 4.1. Processing of ligands. A protein ligand binds to a receptor on the cell surface and is internalized through clathrin-coated pits. After uncoating of the endocytic vesicle, the lumenal pH is reduced by proton pumps in the vesicle membrane. At acid pH the protein (i) may have reduced affinity for its receptor, (ii) have increased affinity for a second receptor, (iii) become a substrate for acidic proteases or (iv) translocate through the endosomal membrane into the cytosol. The events in the endosome may determine the fate of the receptor and its ligand. (Based on ref. 27.)

for *u*ncoupling of *r*eceptors and *l*igands' (CURL) in which a low pH, generated by proton pumps in the membrane, causes dissociation of receptor–ligand complexes.

The intracellular routes followed by different molecules are thus complex. Moreover, recycled products can be released from the cell by exocytosis in the form of small peptides or amino acids, or cross from intracellular vesicles directly into the cytoplasm. In addition, macrophages secrete proteases that can digest the extracellular matrix and other molecules, so that degradation may occur outside the cell. It is unclear which of these pathways, if any, is more relevant to physiological processing of antigens by macrophages before they are presented to T_H cells, but some pertinent observations are now summarized.

2.1.2 Antigen presentation

Ziegler and Unanue (1) examined the ability of macrophages to present bacterial antigens (*Listeria*) to T cells from immunized mice, as measured by their ability to bind and stimulate lymphocyte proliferation in culture. After macrophages had phagocytosed the bacteria, they were unable to present antigens until about 30 min later at 37°C. During this 'lag phase' bacterial antigens were presumably processed, and expressed at the cell surface. Evidence that processing occurred within an acid compartment came from studying the effects of relatively non-specific lysosomotropic agents like ammonium chloride and chloroquine which raises the intracellular pH (2). Macrophages did not present antigens if they were treated with these agents soon after endocytosis, but could if they were treated after the lag phase. Similar results were obtained for T cell hybridomas responding to soluble antigens by Grey and colleagues (3,4).

Shimonkevitz *et al.* (5) found that while chloroquine-treated or glutaraldehyde-fixed (non-viable) macrophages were unable to present native ovalbumin to T cell hybridomas as expected, they could present the denatured molecule or enzyme digests of the antigen. Since the same T cells also responded to native antigen with viable accessory cells, it was reasoned that antigen processing involved a degree of unfolding or degradation of the molecule to expose the immunogenic portions (*Figure 4.2*).

Subsequent studies have shown that T cell epitopes can be defined using synthetic peptides which, in the case of most helper cells, are bound to Ia molecules on the macrophage (see Chapter 1). It is unclear precisely where peptides are produced from exogenous antigens, except that this process may require an acidic compartment (above) and perhaps protease action, because inhibitors like leupeptin can inhibit presentation. Nor is it clear where peptides become associated with Ia molecules, although recent work (6) with human B cell lines indicates one point where the endocytic pathway meets the biosynthetic pathway for MHC molecules. Intracellular transport of class II molecules (which are associated with an 'invariant chain') from the endoplasmic reticulum (ER), where they are synthesized, to the Golgi, where they are sialylated, is a rapid process, but transport from the Golgi to the cell surface is slow. Association of exogenously derived peptides may occur during the latter step, as it has been reported that endocytosis of complexes of neuraminidase with iron-saturated transferrin (in the endocytic pathway) led to removal of sialic acid residues (thus at a post-Golgi stage) from class II MHC β and invariant chains (see *Figure 4.4*).

2.2 B cells

B cells are non-phagocytic but can endocytose molecules by constitutive pinocytosis or after binding to their antigen receptors (membrane Ig).

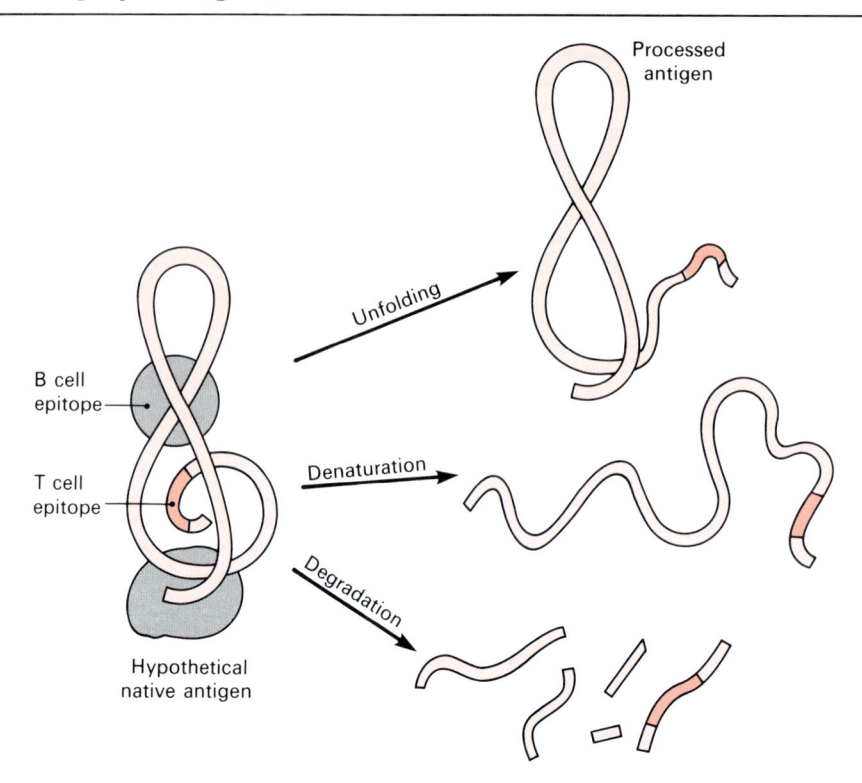

Figure 4.2. Conformational changes that may expose T cell epitopes. Processing of exogenous antigens may require unfolding, denaturation or degradation to allow the T cell epitopes to be exposed and adopt the necessary conformation for binding to an MHC molecule and/or TCR.

2.2.1 Non-specific uptake

B cell lymphomas and B blasts can present antigens to T cell hybridomas. Irrespective of their membrane Ig specificity they can, for example present ovalbumin and keyhole limpet haemocyanin to specific T cells, but, in contrast to macrophages, only a very small amount of intracellular antigen degradation occurs which is not chloroquine-sensitive (3,4).

2.2.2 Uptake via membrane Ig

In antibody responses to hapten-carrier complexes it was thought that B cells bind the hapten via membrane Ig and receive help from T cells that recognize the carrier, probably in association with B cell Ia molecules. This idea has been superseded by the concept that binding to membrane Ig leads to internalization of specific antigens, which are then processed, bound to Ia molecules, and presented to specific T_H cells. In turn, the T cells produce lymphokines and/or other B cell-activating signals (Chapter 2). Thus, although the B cell presents processed antigen to the T cell, it secretes antibodies which generally bind to the native molecule (*Figure 4.3*).

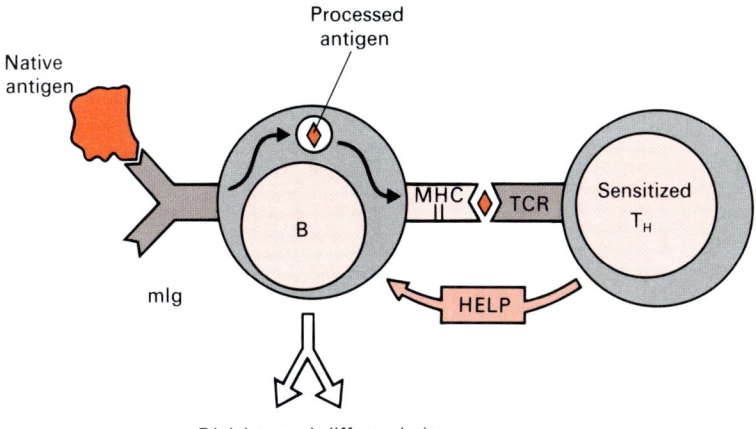

Figure 4.3. Antigen processing and presentation by B cells. B cells take up native antigen via their membrane immunoglobulin (mIg), and can present processed antigen in association with MHC class II molecules to antigen-sensitized T cells.

Evidence for this idea was obtained by Lanzavecchia (7), who used B cell clones specific for tetanus toxoid to present this antigen to T cell clones specific for the same antigen and Ia molecules of the B cell. Antigen-pulsed B cells could stimulate the T cells, but responses were inhibited when antibodies to B cell Ig were added during but not after the pulse, presumably by blocking antigen binding (although inhibitory effects can be generated by antibody binding to B cell Fc receptors: 8). Responses were inhibited by chloroquine treatment or fixation of the B cells during but not after the pulse, suggesting that a pH-dependent process is required, as for processing in macrophages.

2.3 Fibroblasts

For many years it was believed that CTLs recognized native antigens inserted in the plasma membrane of target cells that were somehow associated with class I molecules. There is now increasing evidence in favour of the idea that class I-restricted CTLs, like T_H cells, recognize processed peptides bound to MHC molecules. This also seems likely as cytotoxic and helper T cells can utilize the same pools of TCR V genes (9), and as the structures of class I and class II molecules are similar (10).

Much has been learnt from studies on CTL recognition of influenza, although comparable findings have been made in other systems. Many influenza-specific mouse CTLs, which killed influenza-infected fibroblasts in culture, turned out to be specific for the viral nucleoprotein (NP), even though this molecule was not detectable at the cell surface by antibodies (11). Most transmembrane molecules are glycoproteins that are synthesized on ribosomes associated with the rough ER (RER) and contain a hydrophobic signal peptide which is required

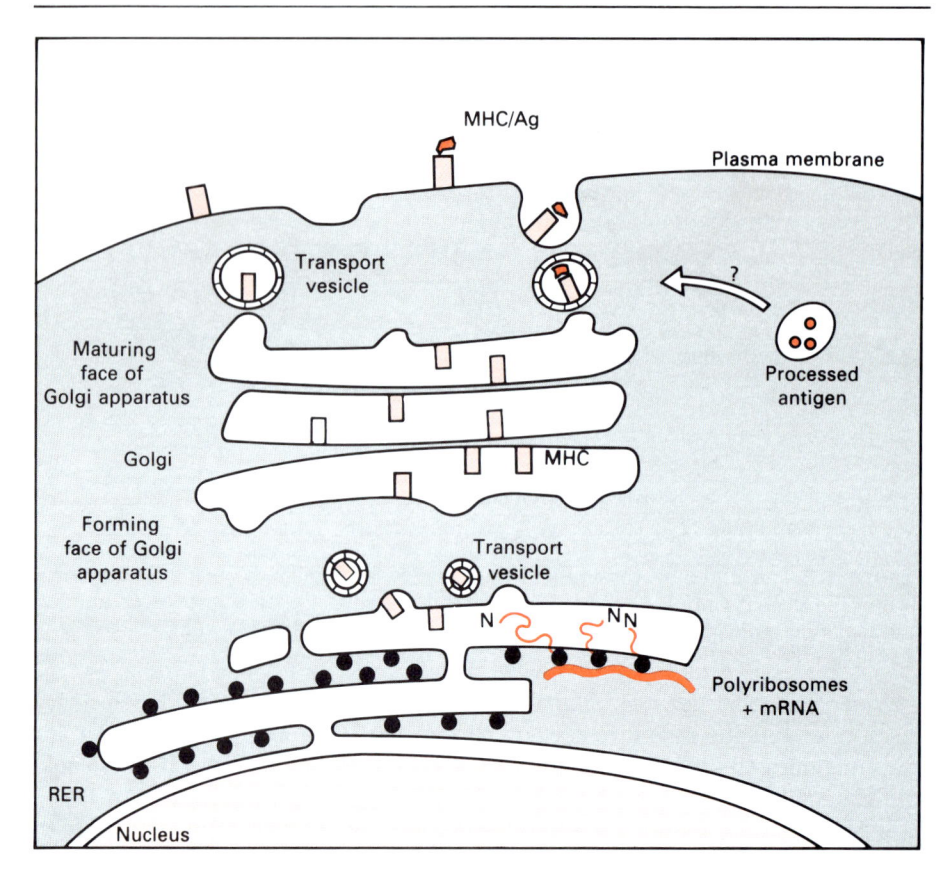

Figure 4.4. Transport of membrane glycoproteins to the cell surface. Polypeptides of (e.g.) MHC molecules are synthesized and inserted into the membrane of the RER from where they are transported in vesicles to the Golgi apparatus. Peptides of processed antigens (Ags) may become associated with MHC molecules within one of these compartments. Membrane molecules, including MHC molecules, are transported from the Golgi to the cell surface in transport vesicles. Based on ref. 28.

for their translocation across the ER, insertion into the membrane and transport to the cell surface (*Figure 4.4*). However, NP is a non-sialylated molecule, that is synthesized on free ribosomes in the cytoplasm and lacks a recognizable signal sequence or transmembrane region, so it is unlikely it could be expressed on the cell surface as an intact molecule. Moreover, another viral component, haemagglutinin (HA), could be recognized by other CTLs even after its signal sequence was deleted and it was expressed in the cytoplasm as a vaccinia construct (12). Therefore pathways for expression of cell surface antigens may be distinct from those for newly synthesized membrane components and may be present in all cells of the body.

Subsequent work showed that CTLs recognize peptides (bound to MHC molecules): uninfected target cells could be pulsed with antigenic peptides and were lysed as efficiently as infected cells (e.g. 13). However, if they were

Table 4.2. Protein degradation via the ubiquitin pathway—the N-end rule[a]

Residue X in ubiquitin – X – protein	Half-life of X – protein
Met Ser Ala Thr Val Gly	>20 h
Ile Glu	~30 min
Tyr Gln	~10 min
Phe Leu Asp Lys	~3 min
Arg	<2 min

[a]From ref. 14.

incubated with native antigens, they were not recognized by class I-restricted CTLs. This has led to the idea that antigens in the cytoplasm of the cell, either synthesized there or having entered from an endosome, are degraded to peptides before presentation. It is not clear how and where this occurs, but as a general rule it does not seem to be a pH-dependent process. Ubiquitin-dependent proteolysis is involved in some cases. This is a pathway by which proteins destined for degradation in the cytosol are modified by attachment of the 8.5 kd polypeptide, ubiquitin. It turns out the half-life of such proteins in the cell (ranging from about 2 min to >20 h) is a function of whichever amino acid is present at the amino-terminal, the so-called N-end rule (14) (*Table 4.2*). By constructing a ubiquitin–NP fusion protein it has been shown that the resultant rapid degradation can overcome defective presentation to CTLs (15). Degradation of foreign antigens may therefore follow similar routes to those for normal turnover of proteins or degradation of aberrant products in the cell.

It is unclear how antigenic peptides become associated with MHC class I molecules. Conceivably this might occur while both are being synthesized and the MHC α chains are folding into their final configuration (a fanciful idea would be that β_2-microglobulin then 'locks' the two in place). This is consistent with data that association of peptides with MHC is quite slow, but once formed the combination is stable (see also Chapter 1). It has been suggested that peptides from exogenous antigens may become associated with class II molecules while those from endogenous antigens bind to class I molecules. For example, class I-restricted HA-specific CTLs lysed targets infected with a vaccinia–HA construct, but not cells incubated with non-infectious virions, whereas class II-restricted CTLs acted conversely (16). This suggested cytoplasmic synthesis was required for presentation to class I-restricted cells, and lysis of virus-infected cells was in fact inhibited by emetine (an inhibitor of protein synthesis) but not by chloroquine. The opposite was found for class II-restricted CTLs. Even so, the pathways may not be so discrete. For example, some soluble proteins or particles that would not be expected to enter the cytoplasm can be presented to class I-restricted CTLs (17).

It is difficult to prove that peptides defined as antigenic *in vitro* are actually presented *in vivo*. An essential requirement seems to be that the molecule has sufficient conformational freedom to expose the relevant determinants (i.e. T cell epitopes) for binding to MHC molecules and/or TCRs. In some cases native antigen needs to be extensively degraded, while in others limited proteolysis may suffice, and sometimes processing may not be required at all. For example, intact fibrinopeptide can be presented to certain T cell clones by fixed accessory cells, apparently because the determinant is in a region of the molecule that already has considerable conformational freedom (18).

3. Mechanisms of immunostimulation

This section focuses on dendritic cells (DCs) as examples of immunostimulatory cells and how they initiate immune responses.

3.1 T cell activation

Dendritic cells are unlike other leukocytes examined because they can activate resting T cells (although we should note that B blasts may have similar activity: Chapter 3). DCs cause initially resting T cells to release interleukin-2 (IL2) and to become responsive to this growth factor, probably by inducing functional IL2 receptors. This was shown in experiments where DCs were cultured with periodate-treated T cells (oxidative mitogenesis); the DCs were killed at various time points, and the ability of the T cells to respond to exogenous IL2 was tested (19). Responsiveness increased with time, and correlated with secretion of IL2 until DNA synthesis was initiated. IL2 stimulates sensitized T cells to proliferate and release other growth factors such as interferon-γ and B cell-activating factors (presumably including IL4 and IL5).

3.2 Clustering

During immune responses in culture, the responding lymphocytes physically associate or 'cluster' with accessory cells (20) (*Figure 4.5*). Sensitized T cells differ in their ability to cluster with DCs and other leukocytes (21). With APCs such as macrophages and B cells, clustering is antigen-dependent and occurs at 37°C and also 4°C. DCs also cluster in this manner but, in addition, they can associate with sensitized lymphoblasts in an antigen-*in*dependent manner at 37°C. Thus, DCs can *cluster* with alloreactive blasts irrespective of their specificity, although subsequent *responses* like lymphokine release are antigen-specific. It is notable that freshly isolated Langerhans cells, which are weakly stimulatory, cannot cluster in this non-specific manner but acquire this ability when they develop into immunostimulatory DCs in culture.

Possibly DCs cluster with resting T cells in a similar antigen-non-specific manner, but the T cells only become activated if they subsequently recognize specific antigen/MHC on the DCs. Otherwise they may dissociate, and more T cells are sampled. The molecules responsible for DC–T cell clustering are not well understood. CD4 and CD8 do not seem to be involved, for antibodies against these molecules do not inhibit clustering even though they inhibit T cell proliferation in the mixed leukocyte reaction (MLR), and the same is true of CD11/18, although specific antibodies against lymphocyte function-associated

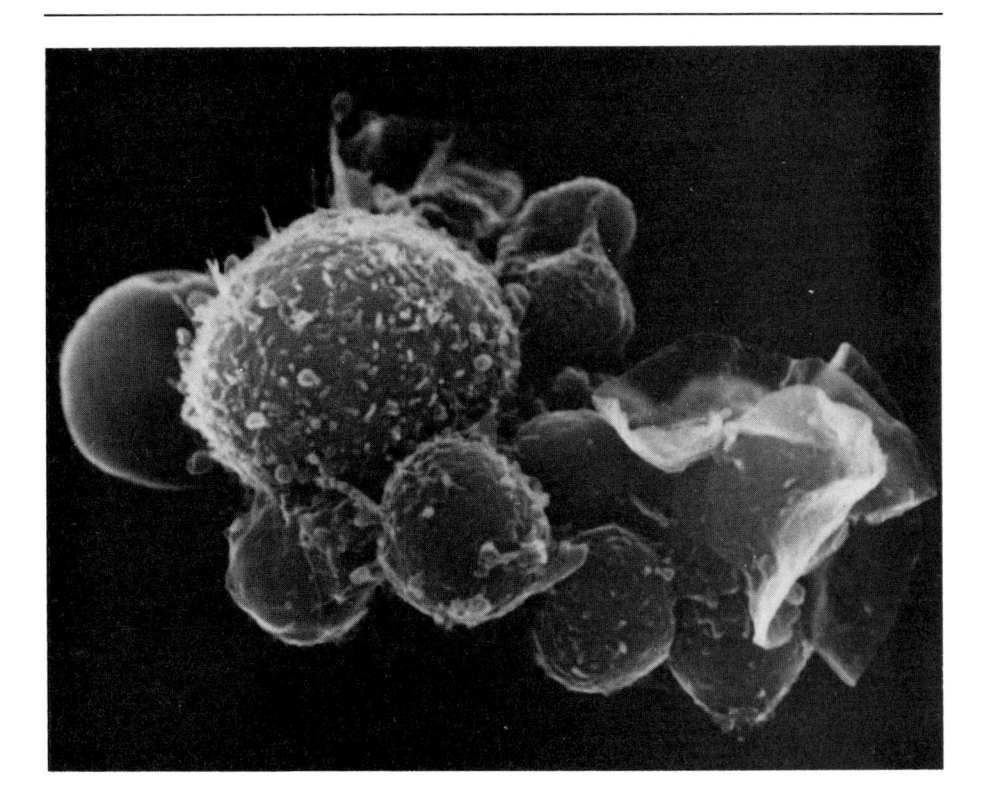

Figure 4.5. Scanning electron micrograph of a cluster of dendritic cells and periodate-treated T cells (in the oxidative mitogenesis response). Note the small round lymphocytes tightly associated with one or more DCs which exhibit lamellipodiae protruding from the cluster. The cluster was isolated before most T cells had begun to blast transform, although a single large, round presumptive lymphoblast is visible. Photograph courtesy of G.MacPherson, from ref. 20.

antigen (LFA-1) do inhibit clustering in oxidative mitogenesis (22). Certainly DC-associated membrane components involved in these interactions have not yet been defined.

3.3 Lymphocyte activating signals

Presumably DCs provide a signal(s) for T cell activation in addition to presenting antigens. It was originally thought that a macrophage-derived cytokine, IL1, was necessary for T cell activation, via an 'interleukin cascade' in which antigen recognition via the TCR, and IL1 binding to the T cell, provided the first and second signals to make the cell responsive to IL2. While IL1 undoubtedly plays a central role in inflammatory responses (23), it is not responsible for T cell activation by DCs, as IL1 is not produced by these cells: no IL1 mRNA or protein was detectable in DCs even after stimulation with lipopolysaccharide or contact with T cells, although it was produced by macrophages (24). However, it acts as a potentiating cytokine by increasing the ability of DCs to cluster with T cells

(25). Preliminary observations (R.M.Steinman, personal communication) also indicate that DCs probably do not synthesize IL6.

What then is the critical DC-derived T cell-activating signal(s)? Possibly DCs secrete other soluble molecules, although no evidence for this has been obtained in double-chamber experiments. Alternatively, they may possess a unique membrane component that interacts wtih a molecule on the T cell, unless the high density of Ia molecules on the DCs or their behaviour in the membrane (e.g. mobility) is important. However, work in progress suggests DCs might form specialized regions of membrane contact with T cells that may prove important for delivery of activation signals (J.M.Austyn and C.Green, unpublished results).

3.4 Antigen presentation by dendritic leukocytes

Dendritic cells can 'present' antigens to resting T cells because the latter are activated in an antigen-specific manner. It seems an anathema that lymphoid DCs, which are so potent at initiating immune responses, are so weakly endocytic, for how, it is argued, can they process and present antigens which they are unable to endocytose? The possibility that DCs have membrane-bound ectoenzymes that process antigens extracellularly has been considered, and to the author's knowledge their presence has not been disproved.

An alternative view comes from more recent observations on the relative capacities of splenic dendritic and Langerhans cells to present antigen (myoglobin) to class II-restricted T cell clones in culture (26). These cells could present specific peptides, but whereas freshly isolated Langerhans cells (LCs) also presented the native antigen, dendritic and cultured Langerhans cells were weak or inactive (*Table 4.3*). This situation represents the converse of their stimulatory activity in the MLR. The results suggest that DC precursors can process and present native molecules (i.e. to sensitized T cells) and then mature into immunostimulatory cells with the capacity to activate antigen-specific,

Table 4.3. Accessory function of freshly isolated and cultured Langerhans cells, splenic macrophages and dendritic cells

Function	Macrophage	Fresh LCs	Cultured LCs	DCs
MHC class II expression	+	+	+	+
Presentation of antigenic peptides to sensitized T cells	+	+	+	+
Processing of native antigen and presentation to sensitized T cells	+	+	−	−
Immunostimulatory activity (e.g. MLR)	−	−	+	+
Antigen-dependent adhesion to sensitized T cells	+	+	+	+
Antigen-independent adhesion to sensitized T cells	−	−	+	+

resting T cells. Presumably LCs, with endocytic activity, internalize native antigen before it is processed and bound to MHC class II molecules, synthesis of which is up-regulated during development in culture.

4. Further reading

Allen,P.M. (1987) Antigen processing at the molecular level. *Immunol. Today,* **9**, 270.
Bodmer,H. and Townsend,A. (1989) Antigen recognition by class I restricted T lymphocytes. *Annu. Rev. Immunol.,* **7**, in press.
Hershko,A. and Ciechanover,A. (1986) The ubiquitin pathway for the degradation of intracellular proteins. *Proc. Nucleic Acid Res. Mol. Biol.,* **33**, 19.
Pernis,B., Silverstein,S.C. and Vogel,H.J. (eds) (1988) *Processing and Presentation of Antigens.* Academic Press Inc., San Diego.
Steinman,R.M., Mellman,I.S., Muller,W.:A. and Cohn,Z.A. (1983) Endocytosis and the recycling of plasma membrane. *J. Cell. Biol.,* **96**, 1.
Unanue,E.R. (1984) Antigen-presenting function of the macrophage. *Annu. Rev. Immunol.,* **2**, 395.

5. References

1. Ziegler,H.K. and Unanue,E.R. (1981) *J. Immunol.,* **127**, 1869.
2. Ziegler,H.K. and Unanue,E.R. (1982) *Proc. Natl. Acad. Sci. USA,* **79**, 175.
3. Chesnut,R.S., Colon,S. and Grey,H.M. (1982) *J. Immunol.,* **129**, 2382.
4. Grey,H.M., Colon.S. and Chesnut,R.S. (1982) *J. Immunol.,* **129**, 2389.
5. Shimonkevitz,R., Kappler,J., Marrack,P. and Grey,H.M. (1983) *J. Exp. Med.,* **158**, 303.
6. Cresswell,P. and Blum,J.S. (1988) In *Processing and Presentation of Antigens.* Pernis,B., Silverstein,S.C. and Vogel,H.J. (eds), Academic Press Inc., San Diego, p. 43.
7. Lanzavecchia,A. (1985) *Nature,* **314**, 537.
8. Bijsterbosch,M.K. and Klaus,G.G.B. (1985) *J. Exp. Med.,* **162**, 1825.
9. Kronenberg,M., Sui,G., Hood,L.E. and Shastri,N. (1986) *Annu. Rev. Immunol.,* **4**, 529.
10. Brown,J.H., Jardetzky,T., Saper,M.A., Samraoui,B., Bjorkman,P.J. and Wiley,D.C. (1988) *Nature,* **332**, 845.
11. Townsend,A.R.M., McMichael,A.J., Carter,N.P., Huddleston,J.A. and Brownlee, G.G. (1984) *Cell,* **39**, 13.
12. Townsend,A.R.M., Bastin,J., Gould,K. and Brownlee,G.G. (1986) *Nature,* **234**, 575.
13. Townsend,A.R.M., Rothbard,J., Gotch,F.M., Bahadur,G., Wraith,D. and McMichael,A.J. (1986) *Cell,* **44**, 959.
14. Bachmair,A., Finley,D. and Varshavsky,A. (1986) *Science,* **234**, 179.
15. Townsend,A., Bastin,J., Gould,K., Brownlee,G., Andrew,M., Coupar,B., Boyle,D., Chan,S. and Smith,G. (1988) *J. Exp. Med.,* **168**, 1211.
16. Morrison,L.A., Lukacher,A.E., Braciale,V.L., Fan,D. and Braciale,T.J. (1986) *J. Exp. Med.,* **163**, 903.
17. Yide Jin,J., Wai-Kuo,S. and Berkower,I. (1988) *J. Exp. Med.,* **168**, 293.
18. Allen,P.M. (1987) *Immunol. Today,* **8**, 270.
19. Austyn,J.M., Steinman,R.M., Weinstein,D.E., Granelli-Piperno,A. and Palladino, M.A. (1983) *J. Exp. Med.,* **157**, 1101.
20. Austyn,J.M., Weinstein,D.E. and Steinman,R.M. (1988) *Immunology,* **63**, 691.

21. Inaba,K. and Steinman,R.M. (1984) *J. Exp. Med.*, **160**, 1717.
22. Austyn,J.M. and Morris,P.J. (1988) *Immunology*, **63**, 537.
23. Kampschmidt,R.F. (1984) *J. Leukocyte Biol.*, **36**, 341.
24. Koide,S.L. and Steinman,R.M. (1987) *Proc. Natl. Acad. Sci. USA*, **84**, 3802.
25. Koide,S.L., Inaba,K. and Steinman,R.M. (1987) *J. Exp. Med.*, **165**, 515.
26. Koide,S., Romani,N., Crowley,M., Witmer-Pack,M., Livingstone,M., Fathman,G.G., Inaba,K. and Steinman,R.M. (1989) *J. Exp. Med.*, submitted.
27. Diment,S., Simmons,B.M., Russell,J.H. and Stahl,P.D. (1988) In *Processing and Presentation of Antigens*. Pernis,B., Silverstein,S.C. and Vogel,H.J. (eds), Academic Press Inc., San Diego, p. 29.
28. Alberts,B., Bray,D., Lewis,J., Raff,M., Roberts,K. and Watson,J.D. (1983) *Molecular Biology of the Cell*. Garland, New York, Chapter 7.

5

Accessory cells *in vivo*

Functions of accessory cells *in vivo* are discussed in this chapter, where we focus on dendritic leukocytes in some primary and secondary lymphoid tissues (Sections 4 and 2) and in non-lymphoid organs (Section 3).

1. Dendritic cell function *in vivo*

Antigens associated with dendritic cells (DCs) are immunogenic *in vivo*, so results obtained in culture indicating that these are immunostimulatory cells (Chapter 3, Section 2) may apply to the whole animal (*Table 5.1*). For example, DCs from draining nodes of mice that were skin-painted with contact sensitizers such as picryl chloride, oxazalone and fluorescein isothiocyanate (FITC) could transfer contact sensitivity against these antigens to naive animals (1,2). In the case of FITC, antigen associated with DCs was visualized by cytofluorography, and labelled cells could also stimulate T cells in an antigen-specific manner in culture. To date, there are very few studies comparing the ability of DCs and other leukocytes to induce antigen responses *in vivo*. However DCs conjugated (*in vitro*) to the hapten trinitrophenol were found to induce contact sensitivity when they

Table 5.1. Stimulation of immune responses by lymphoid dendritic cells *in vivo*

Response	Conditions
Contact sensitivity	Occurs when DCs from draining lymph nodes of animals sensitized to e.g. picryl chloride, oxazolone or FITC are transferred intravenously to naive recipients; when haptenated DCs are administered *in vivo*
Antibody formation	Enhanced anti-viral response when DCs pulsed with virus *in vitro* are administered
Allograft rejection	Triggered when donor strain DCs are administered to recipients before, at the time of, or after transplantation of rat kidney; mouse islets of Langerhans; mouse heart; dGuo-cultured mouse thymus; male skin in female 'non-responder' mice. Host versus graft response also triggered when F1 DCs are injected into parental strain mice, causing splenomegaly/lymph node enlargement

were administered intravenously, whereas similarly coupled macrophages induced tolerance by the same route (3).

Dendritic cells can initiate responses to the 'male-specific' minor histocompatibility antigen, H-Y (4). Female bm12 mice are unresponsive to H-Y, and cannot reject skin grafts from males of the same strain, but this apparent 'Ir (immune response) gene defect' was overcome when females were immunized against male DCs. The ability of DCs to trigger responses to major histocompatibility complex (MHC) antigens is discussed in Section 3. These cells may also be important for (T-dependent) B cell responses *in vivo* (5), as it has been found that antibody responses against tobacco mosaic virus are enhanced by DCs pulsed *in vitro* with the virus.

2. Migration and maturation of dendritic leukocytes

It seems likely that dendritic leukocytes pick up antigens in the periphery (e.g. skin) and migrate to lymphoid tissues (e.g. lymph nodes) to initiate the immune response. When contact allergens were applied to guinea pig skin, veiled cells resembling Langerhans cells were found in cutaneous lymphatics (6), and after skin-painting mice with FITC (Section 1) antigen-labelled DCs accumulated in draining nodes within 8 h and peaked at 24 h (7). The number of dendritic leukocytes in afferent lymph seems to increase in response to antigens and/or inflammatory stimuli. There is also evidence that veiled cells originating from a site(s) in the gut, perhaps the lamina propria and/or Peyer's patches, can transport antigens, because bacterial antigens have been found in veiled cells from *Salmonella*-infected rats (8).

Because Langerhans cells are weakly immunostimulatory (Chapter 4, Section 3.4) they presumably need to mature before they initiate immune responses *in vivo*. Conceivably their entry to lymphatics as veiled cells also occurs after some maturation has occurred. Granulocyte–macrophage colony stimulating factor which may be produced at inflammatory sites causes Langerhans cells to develop into functional DCs in culture, so this cytokine may stimulate their development *in vivo*.

Langerhans cells are long-lived cells that do not proliferate locally in the undisturbed state. For example, if allogeneic skin is transplanted to a nude mouse, the numbers of Langerhans cells remain constant, and host cells replace those of the recipient very slowly (with a half-time of 6 weeks); if the same experiment is done with human skin on a nude mouse, the human cells persist for at least 9 weeks (9). On the other hand, if Langerhans cells are removed from the epidermis by tape-stripping they can be replaced within 1 week (10), showing that a reservoir of precursors can be mobilized rapidly when required.

The distribution of dendritic leukocytes in human skin has been examined during delayed-type hypersensitivity (DTH) responses (11). Langerhans cells (LCs), which are normally found in a suprabasal position in the epidermis, seemed to move upwards and perhaps were eventually sloughed-off. Similar cells then

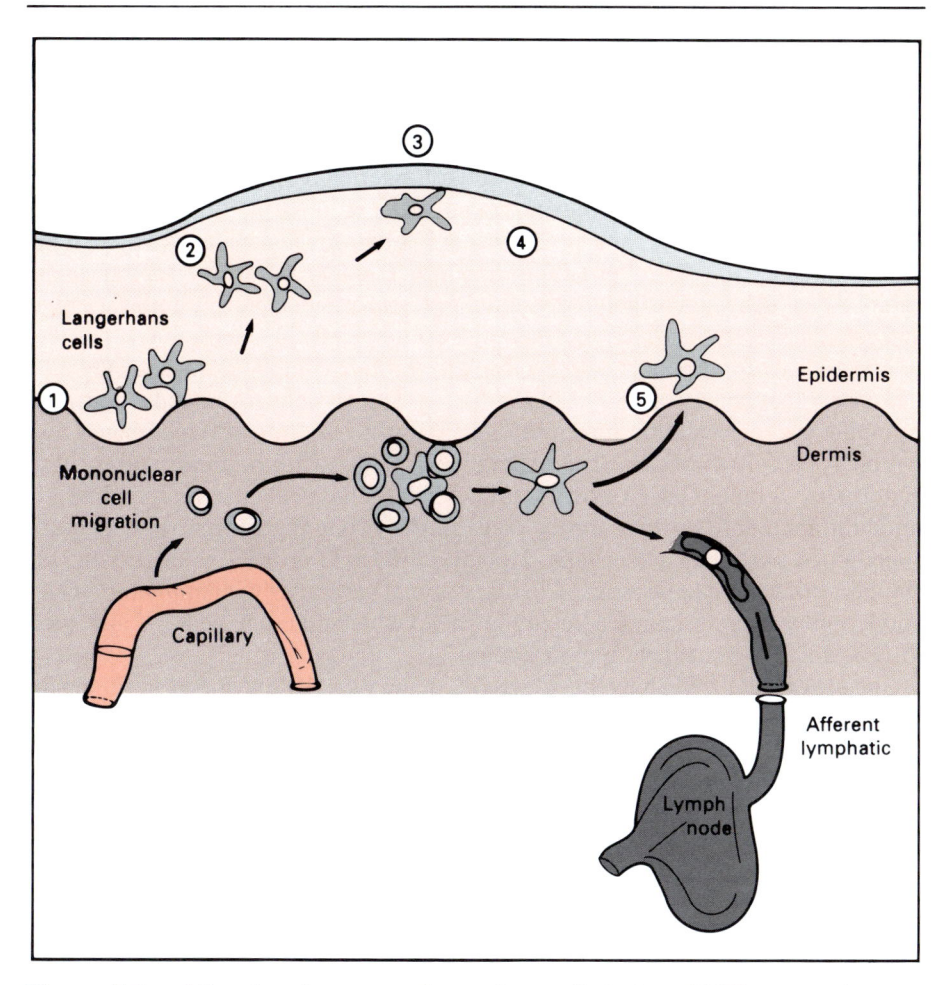

Figure 5.1. Migration of cutaneous Langerhans cells during a DTH response in man. Successive stages are shown left to right. Langerhans cells are initially distributed evenly among keratinocytes (1). Concomitant with keratinocyte hyperplasia and epidermal thickening LCs appear to move upwards (2) and at 72 h they may be sloughed off (3). Within 10 days the epidermis regresses to normal size (4) and is repopulated from the upper dermis (5). In the dermis helper T cells (T_H) and monocytes accumulate perivascularly within 24 h, and dermal LCs appear. Some of these may enter lymphatics as veiled cells while others repopulate the epidermis. (Based on ref. 11.)

appeared in the dermis, and seemed to migrate into the epidermis to replace them. The reason for this is unclear, but conceivably freshly recruited cells can enter the lymphatics as veiled cells (*Figure 5.1*).

2.1 Immunostimulation in lymph nodes

The maturation and migration pathways of dendritic leukocytes may have evolved to ensure that immune responses are initiated in lymphoid tissues rather than the periphery. It seems unlikely that Langerhans cells, lacking stimulatory activity, would be able to initiate responses within the skin itself. These cells may, however, endocytose and process antigens before migrating to draining nodes and acquiring such activity (Chapter 4, Section 3.4). As mature DCs, they may be unable to handle further antigens (at least in native form) but present those carried from the periphery to resting T cells which become activated. This is facilitated by the ability of DCs to home to T cell areas, where they perhaps sample T cells by an antigen-independent clustering mechanism until they find antigen-specific cells which are usually rare (e.g. 10^{-5}).

The ability of lymphocytes to recirculate through lymphoid tissues may also increase the chance of a T cell finding a DC that bears specific antigen. Lymphocyte recirculation refers to their ability to migrate between blood and lymph. The cells first attach to specialized regions within lymph nodes, high endothelial venules (HEV), and cross the endothelial wall into the lymph; their initial attachment (margination) is controlled by 'homing receptors' which are specific for particular endothelia. However, the role of HEV is uncertain, for similar structures can be induced in other areas, particularly inflammatory sites, and lymphocyte recirculation occurs in sheep which do not have HEV and even in fish which do not have lymph nodes.

Once specific T cells have been activated and clonally expanded in the node, sensitized T cells migrate into the periphery. The chances of these T cells finding antigen-presenting (rather than immunostimulatory) cells (APCs) is now increased because there are more of them. Activated T cells bind more strongly to peripheral endothelium than resting T cells, particularly at sites of inflammation. This is related to the induction of adhesion molecules on endothelium by cytokines and to the presence of additional receptors for them on activated T cells. Conversely, resting T cells bind preferentially to HEV in lymphoid tissues. Thus the migratory properties of T cells fits with the concept of central immunostimulation, division and differentiation and finally interaction with APCs in the tissues. When sensitized T_H cells recognize peripheral APCs, such as macrophages, the latter are converted into more efficient effector cells that can eliminate antigens (e.g. bacteria and viruses) locally. These ideas have important implications for transplantation reactions, discussed below.

3. Dendritic leukocytes in transplantation

3.1 Stimulation of allograft rejection

The fact that dendritic leukocytes are present in non-lymphoid organs, together with their role in initiating immune responses, suggests they may trigger allograft rejection. Such cells were called 'passenger leukocytes' (12) because they were originally thought to be carried over to the host as passengers in the vasculature of transplanted organs (13). An alternative view is that these cells are actually 'resident' DCs in non-lymphoid tissues. The bulk of an allograft, though foreign,

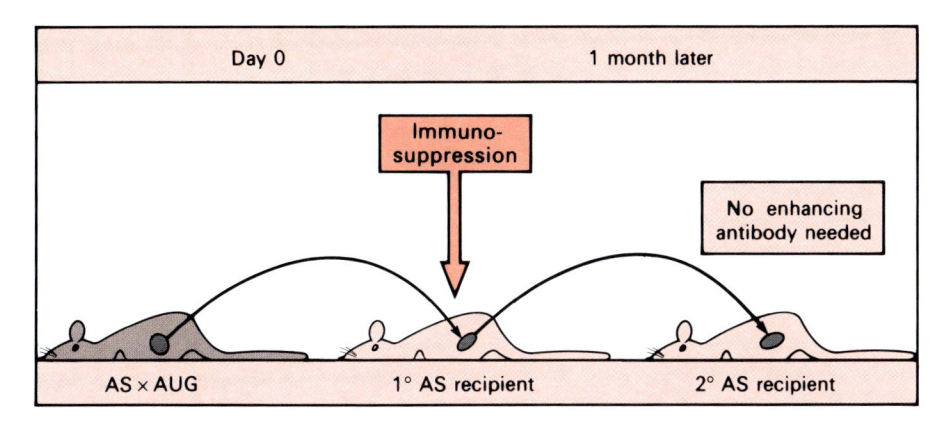

Figure 5.2. Role of passenger leukocytes in graft rejection. A semi-allogeneic kidney is transplanted from an (AS × AUG)F1 rat into an AS recipient. To enable this graft to be accepted the recipient must be immunosuppressed. When this kidney is retransplanted into a second AS rat 1 month later the graft is accepted without immunosuppression (enhancing antibody).

may not be recognized until the host has been sensitized by passenger leukocytes, so if the latter could be removed or inactivated before transplantation, graft acceptance might be possible with less (sometimes life-threatening) immunosuppression of the recipient.

There is evidence that DCs trigger graft rejection *in situ*, and that they may be important passenger leukocytes. For example, rat kidneys that were parked in allogeneic recipients (whose normal immune response was overcome experimentally) were not rejected when they were retransplanted to naive recipients of the same strain as the latter, even though the bulk of the graft expressed allo-antigens (14) (*Figure 5.2*). It was proposed that donor passenger leukocytes were replaced by those of the intermediate recipient which were not immunogenic (since they were the same strain as the secondary recipient). Rejection was triggered when small numbers of donor-strain DCs (but not other leukocytes) were administered at the time of retransplantation.

More direct evidence was obtained in the mouse, where treatment of pancreatic islets of Langerhans with the anti-DC antibody 33D1 and complement *in vitro* resulted in prolonged survival in allogeneic recipients, but donor-strain DCs given later triggered graft rejection (15). Pre-treatment with allogeneic DCs can also sensitize recipients before transplantation, resulting in accelerated rejection of cardiac allografts in the mouse (16), and rejection of embryonic thymus lobes that had been cultured in deoxyguanosine to kill DCs and which are normally accepted by allogeneic recipients (17) (Section 4). Culture of various tissues (thyroid, parathyroid and islets of Langerhans) under other conditions, particularly in high concentrations of O_2 also reduces their immunogenicity, apparently because passenger leukocytes are killed (18,19).

A major problem is that these strategies are only effective in limited strain combinations. Another is that lymphoid DCs and veiled cells have been tested

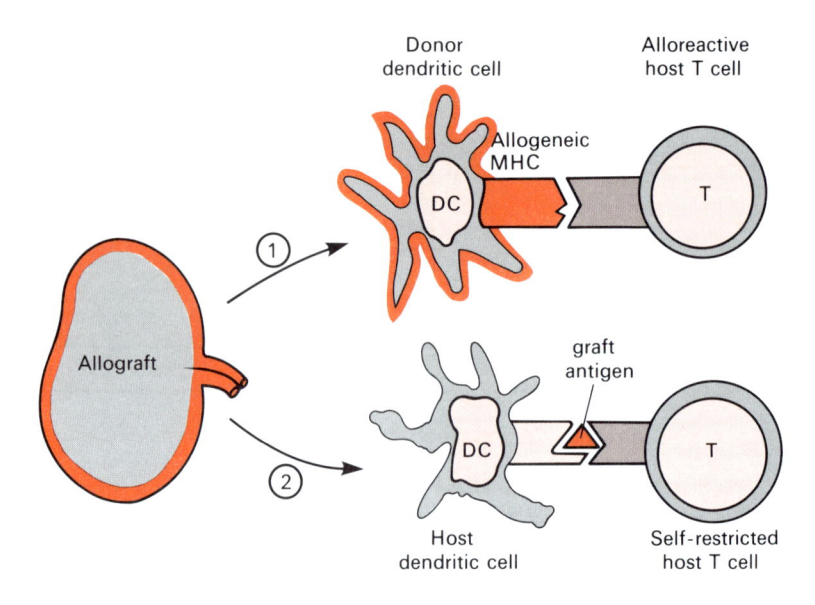

Figure 5.3. Routes for sensitization of recipients to graft antigens. (**Route 1**) Dendritic cells from the graft present their own MHC molecules to alloreactive T cells. (**Route 2**) Host dendritic cells present processed graft antigens to self-restricted T cells.

for their ability to cause rejection, but the activity of dendritic leukocytes from non-lymphoid tissues has not yet been evaluated. Currently it is thought that recipient T cells can be sensitized directly by graft DC that bear allo-antigens ('route 1') or by host DCs that present graft antigens ('route 2'); where this might occur is considered in Section 3.2 (*Figure 5.3*).

3.2 Central versus peripheral sensitization
The idea that antigen responses are induced 'centrally' in lymphoid tissues such as lymph nodes (Section 2.1) is compatible with early work on skin graft rejection. Skin allografts placed on alymphatic pedicles (skin flaps separated from the host by a plastic dish that interrupted the lymphatics: *Figure 5.4*) enjoyed greatly prolonged or indefinite survival, and graft rejection only occurred when lymphatic connections were re-established (20). One interpretation would be that allogeneic Langerhans cells normally migrate from the skin graft to draining nodes and stimulate alloreactive T cells, but their migration was inhibited by interrupting the lymphatics. The importance of lymphatics in allograft rejection was also suggested by the prolonged survival of allografts placed in 'privileged sites' (e.g. brain, anterior chamber of the eye, hamster cheek pouch) ostensibly because these areas lacked or had unusual lymphatic drainage (21).

In contrast to skin allografts, however, rejection of fully vascularized allografts (i.e. directly 'plumbed-into' the circulation) did not seem to require an intact

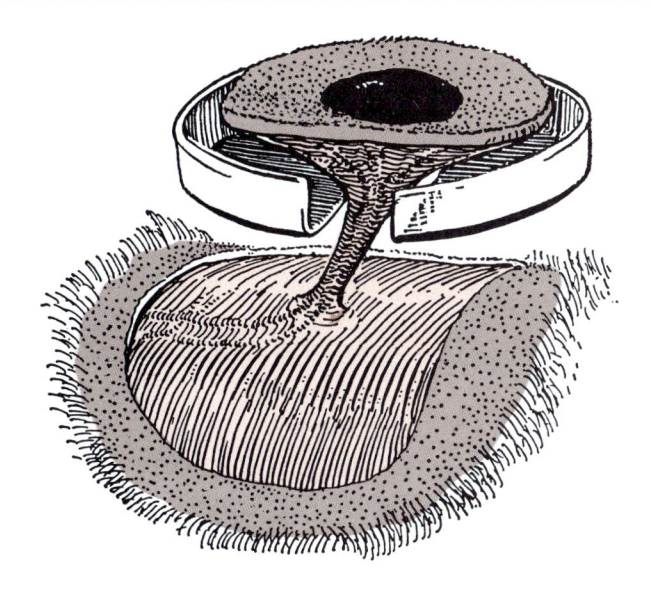

Figure 5.4. Role of lymphatics in graft rejection. Alymphatic pedicles were prepared by cutting skin flaps from the flanks of anaesthetized guinea pigs, but retaining a vascular 'umbilical cord'. The flap was isolated from the underlying skin by a plastic dish. When an allograft (black circular area) was inlaid on the host skin it survived for a prolonged period or indefinitely due to interruption of the lymphatic drainage. (Based on ref. 20.)

lymphatic drainage (22). Thus, kidney allografts placed in plastic envelopes or externally to the recipient were rejected normally. Such observations led Medawar to propose the idea of 'peripheral sensitization', whereby T cells become sensitized within the graft itself (21). An alternative possibility, that does not seem to have been entertained at the time, is that sensitization actually occurred centrally in the spleen, by way of the blood.

3.2.1 Migration of DC into the spleen

Dendritic cells can migrate from the blood to T areas of spleen, by a route that seems to parallel traffic via lymphatics into the node (23). When lymphoid DCs were purified from mouse spleen, labelled with a radioisotope (indium-111) and injected intravenously, they were initially retained in the lungs but soon migrated to other organs. The spleen turned out to be the primary site of DC localization (on the basis of radioactivity per weight of tissue) and, except for the liver, DCs were not detectable in other tissues including lymph nodes. DCs were then labelled with a blue fluorochrome to visualize where they were localized in the spleen by double-labelling frozen sections with green-fluorescing antibodies (e.g. anti-T cell) (24). Three hours afer injection, DCs were in the red pulp but at 24 h they were in T cell areas. Therefore mature, immunostimulatory DCs can migrate from the blood (perhaps carrying antigens in physiological situations) into the spleen and probably become interdigitating cells (IDCs).

Traffic of mature DCs into the spleen seems to be controlled by T cells, because DCs do not enter spleens of nude mice unless they are reconstituted with T cells (23). Possibly T cells modify the endothelium in splenic marginal zone to allow recruitment of DCs from the blood, and live DCs attach specifically to this region if they are incubated on frozen sections of normal (euthymic) spleen (24). (Parenthetically, this cannot be the whole story for nude spleens do in fact contain IDCs.) These observations raised the possibility that T cells accumulating in other sites (e.g. during inflammation) may be able to recruit immunostimulatory DCs from the blood, so that immune responses might be generated in the periphery (i.e. peripheral sensitization). This was examined in the context of allograft rejection and found *not* to occur (Section 3.2.2).

3.2.2 Migration of DCs during graft rejection

During rejection of rat cardiac allografts, donor-strain DCs were reported in clusters with host T cells in the heart (25), a finding perhaps consistent with peripheral sensitization ('route 1': Section 3.1), although they were only evident 3–4 days after transplantation and might have been clusters of graft leukocytes with host T cells that were already activated. However, it was shown by Sherwood *et al.* (26) that host accessory cells, perhaps DCs, can present graft antigens to host T cells and trigger graft rejection ('route 2': Section 3.1).

We reasoned that if sensitization by host DCs occurs peripherally one might detect traffic of DCs into allografts, so the migration of intravenously injected radiolabelled DCs was traced in mice bearing fully vascularized heterotopic cardiac allografts and skin allografts (C.P.Larsen, H.Barker, P.J.Morris and J.M.Austyn, in preparation). In fact there is *no* detectable migration of mature host-strain DCs into allografts, certainly not at early times after transplantation when the animal becomes sensitized (instead, the DCs behaved as in normal animals). Therefore, if host DCs do sensitize host T cells peripherally, they must enter allografts as precursors.

Currently, however, we favour the idea that sensitization occurs centrally, after transplantation of vascularized organs as well as skin grafts. We (C.P.Larsen and J.M.Austyn, unpublished results) have observed that soon after cardiac transplantation in the mouse, cells of graft origin, perhaps DCs, can be found in host spleen presumably migrating by way of the blood. If this is correct, there could be parallels with Langerhans cells and migration via the lymphatics. Dendritic leukocytes in non-lymphoid tissues may be immature (i.e. not immunostimulatory) and need to mature before they can migrate via the blood into the spleen and initiate immune responses. In support of this, we have some data that dendritic leukocytes from mouse kidneys are precursors that lack immunostimulatory function, but develop into DCs in culture (J.M.Austyn, unpublished results). Physiologically, this means the function of dendritic leukocytes in non-lymphoid tissues (including skin) may be to sample their environment and migrate to lymphoid tissues to initiate immune responses if foreign antigens are present.

The migration of sensitized lymphoblasts was also examined in parallel experiments to those with DCs in animals bearing cardiac allografts. In contrast to DCs, these cells entered allografts but, as was found by others, in a non-specific manner since host T cells sensitized to graft or to irrelevant 'third-party' antigens entered grafts similarly. It presumably makes biological sense for sensitized T cells to be recruited to any inflammatory site, for there is no way of 'knowing' beforehand whether the specific antigen will be encountered on an APC at a given site or not.

4. Dendritic cells in the thymus

4.1 T cell development

The thymus has three distinct layers: the subcapsular region, the cortex, and innermost, the medulla. Within these regions, T cells at various stages of

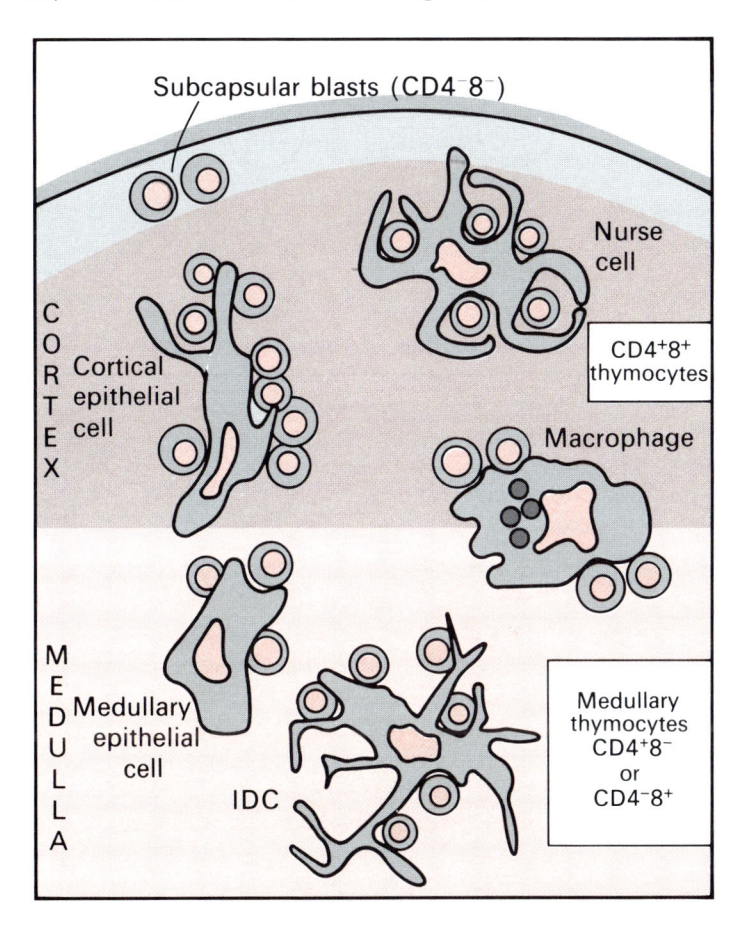

Figure 5.5. Schematic diagram of cell types in mouse thymus, based on ref. 46. Thymocytes are seen to cluster around different types of potential APCs.

68 Antigen-presenting cells

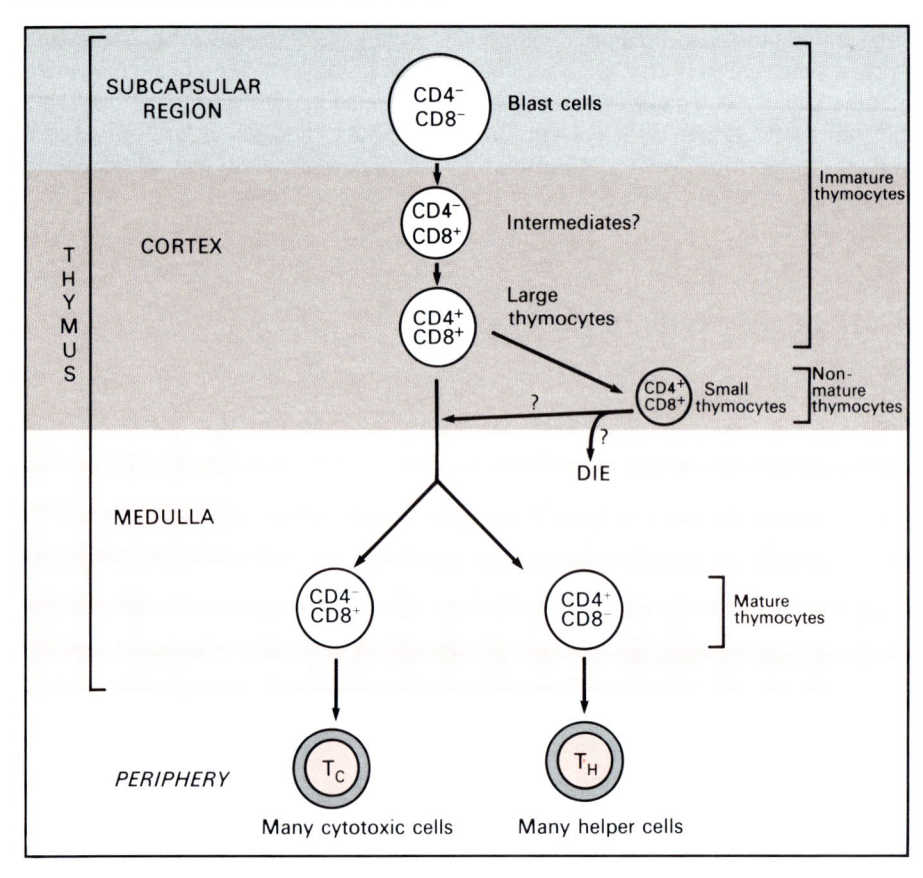

Figure 5.6. T cell development in thymus. This simplified scheme is based on the expression of the CD4 and CD8 markers only. The subcapsular and cortical subsets are heterogeneous for other markers such as the interleukin-2 (IL2) receptor, J11d, CD5, MHC class I and peanut agglutinin (47).

development are associated with different non-lymphoid cells (*Figure 5.5*). In the subcapsular region, large proliferating 'double negative' (CD4⁻CD8⁻) lymphoblasts, which are probably the earliest T cell precursors, are found together with thymic nurse cells. Within the cortex there are mostly small, double positive (CD4⁺CD8⁺) cells, comprising about 90% of the total thymocyte population, in association with a ramifying network of cortical epithelial cells. Macrophages are localized particularly in the cortico-medullary region, while single positive (CD4⁺CD8⁻ or CD4⁻CD8⁺) thymocytes, thought to be the most mature cells, are localized in the medulla in association with medullary epithelial cells and IDCs. The latter are bone marrow-derived cells that closely resemble IDCs in secondary lymphoid tissues (27) and antigens have access to these cells from the circulation (28).

One view holds that T cell development proceeds from the outermost subcapsular region towards the medulla (*Figure 5.6*), although this has not been proved and is vigorously disputed (29,30). The functions of the different types of non-lymphoid cell in the thymus are poorly understood, but there is some information on possible contributions of epithelial cells and DCs in T cell development, the focus of this section. Briefly, it is proposed that epithelial cells determine self-restriction, whereas DCs are responsible for self-tolerance in developing T cells. Some evidence for and against this is outlined below.

4.2 *The restriction repertoire*

The 'restriction repertoire' refers to which set (or haplotype) of MHC molecules is used during antigen recognition by T cells. It turns out that many T cells are self-restricted (Chapter 1, Section 3.4) and preferentially seem to recognize foreign antigens bound to MHC molecules that were encountered during their development (31). Normally, T cells encounter and become restricted to self-MHC molecules, but after an allogeneic bone marrow transplant the T cells become restricted to the MHC molecules of their new environment. For example, T cells from F1 animals [e.g. (C × D) F1, where C and D are MHC genotypes] can recognize foreign antigens bound to either C or D strain MHC molecules. But if F1 bone marrow is transplanted to an irradiated C strain animal, whose own lymphocytes and stem cells have thus been destroyed, T cells that develop in this 'chimera' recognize antigens bound to C strain but *not* D strain MHC molecules, and T_H cells from these animals only proliferate in response to APCs, or help B lymphocytes, from strain C (*Figure 5.7*). Similar results are obtained if a C strain thymus is transplanted to an F1 nude mouse.

These and other experiments showed that the restriction of T_H cells was

Experiment	C × D Bone marrow cells → C → (X) Repopulation of immune system & maturation → C	
Genotype of thymus	C × D	C
Genotype of developing T cells	C × D	C × D
T cells restricted to MHC — D	+	−
T cells restricted to MHC — C	+	+

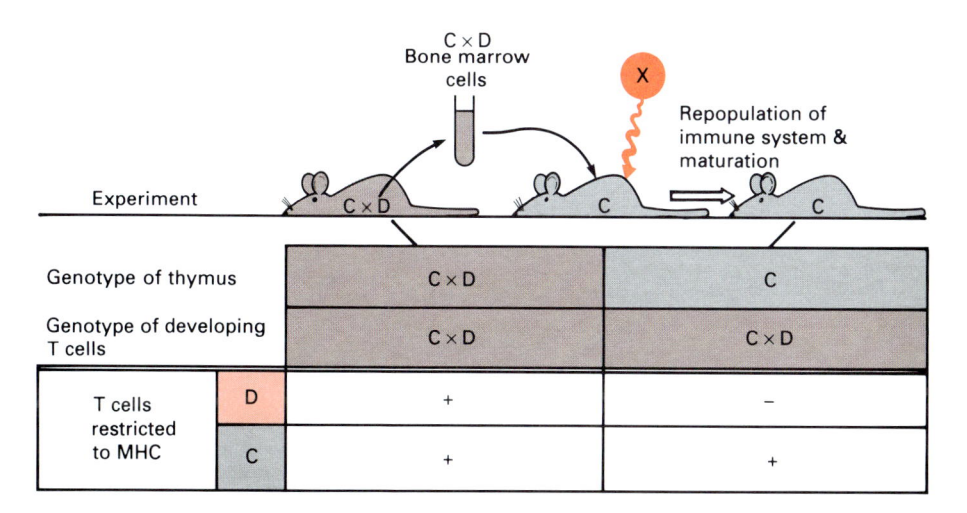

Figure 5.7. Development of T cell MHC restriction. Bone marrow cells from a mouse of MHC haplotype (C × D) were used to repopulate an irradiated (X) type C recipient (top). Splenocytes (mature lymphocytes) from a (C × D) animal are normally restricted to (i.e. recognize antigens on) APCs of type C or D. However (C × D) splenocytes which matured in the type C animal, produced as above, are restricted to APCs of type C only.

70 Antigen-presenting cells

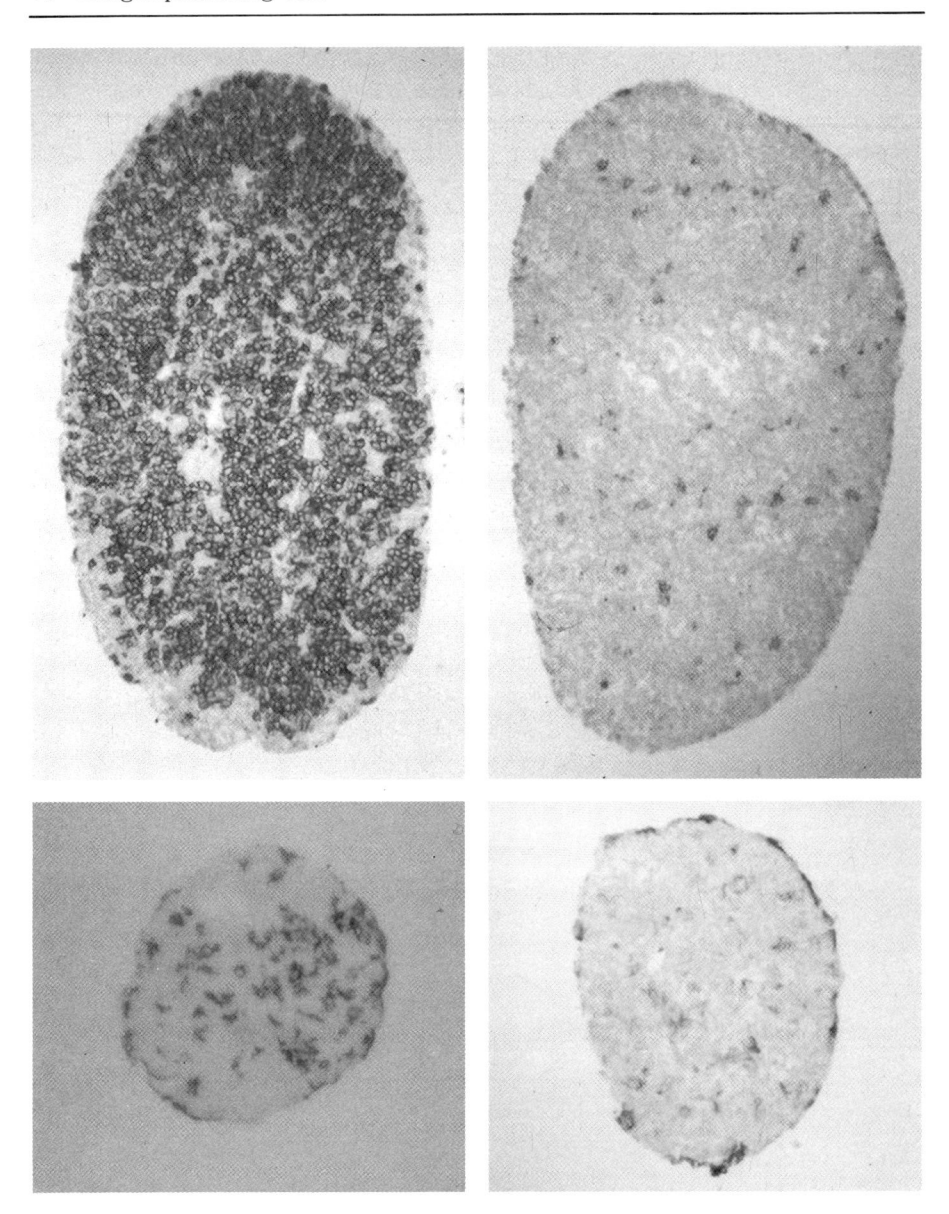

Figure 5.8. Normal and deoxyguanosine-cultured embryonic thymus lobes. Thymus lobes were dissected from 14 day mouse embryos and cultured for several days in the absence (top) or presence (bottom) of 1.35 mM dGuo. Serial sections of the two lobes were then stained with antibodies specific for the leukocyte common antigen (left) or the F4/80 macrophage antigen (right). The lobe cultured without dGuo has increased in size and contains many leukocytes (top left), most of which are T cells and precursors (not shown) with some macrophages (top right). After culture in dGuo, the lobe is small and contains very few leukocytes (bottom left) most of which seem to be macrophages (bottom right), but there are few if any thymocytes. Thymic dendritic cells are probably also killed by dGuo (not shown). Courtesy of P.Fairchild.

determined by a relatively radioresistant cell in the thymus although the restriction of cytotoxic T cells can be determined extrathymically (e.g. 32). Longo and Schwartz (33) presented evidence that the 'thymic restricting cell' was bone marrow-derived. Although F1 T cells that developed in a C strain environment (discussed above) were initially restricted to C, T cells produced later were restricted to both C and D. Presumably the original F1 thymic restricting cells had died and been replaced with C strain cells from the marrow; candidate cells would be thymic IDCs, which unlike DCs in the periphery may be relatively radioresistant and long-lived. This was contested by Lo and Sprent (34) who argued instead that restriction was imparted by thymic epithelial cells (which do not originate in the marrow). Thymus lobes were cultured in deoxyguanosine (dGuo), which depletes DCs and thymocytes to leave the epithelial framework (35) (*Figure 5.8*). When dGuo-treated lobes from C strain animals, were transplanted to F1 ATXBM mice (Chapter 2, Section 3) the F1 T cells became restricted to C, suggesting that thymic epithelium was sufficient and IDCs may not be important in determining restriction. Other evidence also points to the same conclusion (36).

4.3 Self-tolerance

The mechanism by which T cells become tolerant to self components appears to be different from that determining restriction. Put simply, merely being tolerant to a particular set of MHC molecules (with or without bound peptides) does not result in the ability to recognize foreign antigens bound to those molecules. For example, F1 T cells that develop in a C environment (Section 4.2) are restricted to C, but tolerant to both C and D; the same applies for 'double-donor' chimeras with C strain and D strain marrow in an irradiated C recipient. Nor are T cells from animals made neonatally tolerant to particular MHC molecules restricted to them.

It has become clear over the past year or two that T cell tolerance can result from clonal deletion. The first observation was that T cells expressing a particular variable element, $V_\beta 17a$, as detected by a monoclonal antibody, were deleted in strains of mice expressing I-E, apparently because this element conferred reactivity against (self) I-E (37). This mechanism now seems the rule in other cases (38), although the extent to which deletion occurs at the level of thymic double-positive cells differs perhaps depending on whether class I or class II restriction is involved. In contrast, other mechanisms can be responsible for B cell tolerance (e.g. 39).

To compare the role of thymic epithelium and other cells in the development of tolerance, use was again made of dGuo-cultured thymic lobes (Section 4.2). When these were transplanted to allogeneic recipients, they were not rejected presumably because the passenger leukocytes were depleted (Section 3.1), but the recipients did not become tolerant to the donor MHC molecules as determined by proliferation in the MLR (40), nor did CTLs (41). This suggests dGuo-sensitive cells, perhaps IDCs rather than epithelial cells, determine tolerance. However, the opposite result was obtained (42) if lobes were cultured at 24°C before transplantation to nude animals, but the extent to which these lobes were fully depleted can be questioned. Even so, tolerance of CTLs to minor histocompatibility antigens apparently did develop after grafting dGuo lobes (43). This may be due to different effects of the thymus on helper and cytotoxic T cell development but more recent experiments have shown that CTLs that matured in the presence of allogeneic DCs became tolerant to MHC molecules of the latter (44).

In summary, there are no clear cut answers, but one current working hypothesis is that T cells bearing receptors with a wide range of affinities for self-MHC may be positively selected by interaction with epithelial cells, while those with moderate affinity are subsequently deleted when they encounter bone marrow-derived cells, possibly thymic IDCs. Alternatively, positive selection may not occur in this manner, but bone marrow-derived cells do determine tolerance (45).

5. Further reading

Adkins,B., Mueller,C., Okada,C.Y., Reichert,R.A., Weissman,I.L. and Spangrude,G.J. (1987) Early events in T-cell maturation. *Annu. Rev. Immunol.*, **5**, 325.
Austyn,J.M. and Steinman,R.M. (1988) The passenger leukocyte—a fresh look. *Transplant Rev.*, **2**, 139.
Butcher,E.C. (1986) The regulation of lymphocyte traffic. *Curr. Topics Microbiol. Immunol.*, **128**, 85.
Butcher,E.C. and Weissman,I.L. (1984) Lymphoid tissues and organs. In *Fundamental Immunology*. Paul,W.E. (ed.), Raven Press, New York, p. 109.
Von Boehmer,H. (1988) The developmental biology of T lymphocytes. *Annu. Rev. Immunol.*, **6**, 309.

6. References

1. Knight,S.C., Krejci,J., Malkovsky,M., Colizzi,V., Gautam,A. and Asherson,G.L. (1985) *Cell. Immunol.*, **94**, 427.
2. Macatonia,S.E., Edwards,A.J. and Knight,S.C. (1986) *Immunology*, **59**, 509.
3. Britz,J.S., Askenase,P.W., Ptak,W., Steinman,R.M. and Gershon,R.K. (1982) *J. Exp. Med.*, **155**, 1344.
4. Boog,C.J.P., Kast,W.M., Timmers,H.Th.M., Boes,J., De Waal,L.P. and Melief, C.J.M. (1985) *Nature*, **318**, 59.
5. Francotte,M. and Urbain,J. (1985) *Proc. Natl. Acad. Sci. USA*, **82**, 8149.
6. Silberberg,I.R., Baer,L. and Rosenthal,S.A. (1976) *J. Invest. Dermatol.*, **66**, 210.
7. Macatonia,S.E., Knight,S.C., Edwards,A.J., Griffiths,S. and Fryer,P. (1987) *J. Exp. Med.*, **166**, 1654.
8. Mayrhofer,P., Holt,P.G. and Papadimitriou,J.M. (1986) *Immunology*, **58**, 379.
9. Krueger,G.G., Daynes,R.A. and Mansoor,E. (1983) *Proc. Natl. Acad. Sci. USA*, **80**, 1650.
10. Streilein,J.W., Lonsberry,L.W. and Bergstresser,P.R. (1982) *J. Exp. Med.*, **155**, 863.
11. Kaplan,G., Nusrat,A., Witmer,M.D., Nath,I. and Cohn,Z.A. (1987) *J. Exp. Med.*, **165**, 763.
12. Elkins,W.L. and Guttmann,R.D. (1968) *Science*, **159**, 1250.
13. Snell,G.D. (1957) *Annu. Rev. Microbiol.*, **11**, 439.

14. Lechler,R.I. and Batchelor,J.R. (1982) *J. Exp. Med.*, **155**, 31.
15. Faustman,D., Steinman,R.M., Gebel,H., Hauptfeld,V., Davis,J. and Lacy,P. (1984) *Proc. Natl. Acad. Sci. USA*, **81**, 3864.
16. Peugh,W.N.P., Austyn,J.M., Carter,N.P., Wood,K.J. and Morris,P.J. (1987) *Transplantation*, **44**, 706.
17. Jenkinson,E.J., Benson,M.T., Buckley,G. and Owen,J.J.T. (1987) *Immunology*, **60**, 593.
18. Lafferty,K.J., Prowse,S.J. and Simeonovic,C.J. (1983) *Annu. Rev. Immunol.*, **1**, 143.
19. Lacy,P.E. and Davie,J.M. (1984) *Annu. Rev. Immunol.*, **2**, 183.
20. Barker,C.F. and Billingham,R.E. (1968) *J. Exp. Med.*, **128**, 197.
21. Billingham,R. and Silvers,W. (1971) *The Immunobiology of Transplantation*. Osler,A.G. and Weiss,L. (eds), Foundations of Immunology Series, Prentice-Hall, Inc.
22. Hume,D.M. and Egdahl,R.H. (1955) *Surgery*, **38**, 194.
23. Kupiec-Weglinski,J.W., Austyn,J.M. and Morris,P.J. (1988) *J. Exp. Med.*, **167**, 632.
24. Austyn,J.M., Kupiec-Weglinski,J.W., Hankins,D.F. and Morris,P.J. (1988) *J. Exp. Med.*, **167**, 646.
25. Forbes,R.D.C., Parfrey,N.A., Gomersall,M., Darden,A.G. and Guttmann,R.D. (1986) *J. Exp. Med.*, **164**, 1239.
26. Sherwood,R.A., Brent,L. and Rayfield,L.S. (1986) *Eur. J. Immunol.*, **16**, 569.
27. Barclay,N.A. and Mayrhofer,G. (1981) *J. Exp. Med.*, **153**, 1666.
28. Kyewski,B.A., Fathman,C.G. and Rouse,R.V. (1986) *J. Exp. Med.*, **163**, 231.
29. Reichert,R.A., Gallatin,W.M., Butcher,E.C. and Weissman,I.L. (1984) *Cell*, **38**, 89.
30. Shortman,K., Mandel,T., Andrews,P. and Scollay,R. (1985) *Cell. Immunol.*, **93**, 350.
31. Schwartz,R.H. (1984) In *Fundamental Immunology*. Paul,W.E. (ed.), Raven Press, New York, p. 379.
32. Bradley,S.M., Kruisbeek,A.M. and Singer,A. (1982) *J. Exp. Med.*, **156**, 1650.
33. Longo,D.L. and Schwartz,R.H. (1980) *Nature*, **287**, 44.
34. Lo,D. and Sprent,J. (1986) *Nature*, **319**, 672.
35. Jenkinson,E.J., Franchi,L.L., Kingston,R. and Owen,J.J.T. (1982) *Eur. J. Immunol.*, **12**, 583.
36. Ron,Y., Lo,D. and Sprent,J. (1986) *J. Immunol.*, **137**, 1764.
37. Kappler,J.W., Roehm,N. and Marrack,P. (1987) *Cell*, **49**, 273.
38. Pullen,A.M., Marrack,P. and Kappler,J.W. (1988) *Nature*, **335**, 796.
39. Goodnow,C.C., Crosbie,J., Adelstein,S., Lavoie,T.B., Smith-Gill,S.J., Brink,R.A., Pritchard-Briscoe,H., Wotherspoon,J.S., Loblay,R.H., Raphael,K., Trent,R.J. and Basten,A. (1988) *Nature*, **334**, 676.
40. Ready,A.R., Jenkinson,E.J., Kingston,R. and Owen,J.J.T. (1984) *Nature*, **310**, 231.
41. Von Boehmer,H. and Schubiger,K. (1984) *Eur. J. Immunol.*, **14**, 1048.
42. Jordan,R.K., Robinson,J.H., Hopkinson,N.A., House,K.C. and Bentley,A.L. (1985) *Nature*, **314**, 454.
43. Von Boehmer,H. and Hafen,K. (1986) *Nature*, **320**, 626.
44. Matzinger,P. and Guerder,S. (1989) *Nature*, **338**, 74.
45. Marrack,P., Lo,D., Brinster,R., Palmiter,R., Burkly,L., Flavell,R.H. and Kappler,J. (1988) *Cell*, **53**, 627.
46. Butcher,E.C. and Weissman,I.L. (1984) In *Funbamental Immunology*. Paul,W.E. (ed.), Raven Press, New York, p. 109.
47. Adkins,B., Mueller,C., Okada,C.Y., Reichert,R.A., Weissman,I.L. and Spangrude, G.J. (1987) *Annu. Rev. Immunol.*, **5**, 325.

Glossary

Accessory cell: any cell that, together with antigen, stimulates a lymphocyte response. The term can be used collectively for antigen-presenting cells and immunostimulatory cells.

Antigen: a molecule recognized by the immune system.

Alloreactivity: the ability of T cells from an individual to recognize allogeneic MHC molecules, with or without bound peptides, from another member of the species (also referred to as allo-antigen). Alloreactive T cells may respond to foreign antigens bound to self-MHC molecules in physiological situations (i.e. cross-reactive recognition).

Allografts: tissues transplanted from unrelated members of the species (originally called homografts).

Antigen-presenting cells (APCs): cells that present foreign antigens to T cells; normally, such antigens are bound to MHC molecules on the APC. Originally the term APC was applied particularly to cells expressing MHC class II molecules, but should now be used for any cell that presents antigen to T cells regardless of which class of MHC molecules it possesses.

Antigen processing: conversion of a native antigen (e.g. globular protein in its normal, folded configuration) to forms such as peptides that can bind to MHC molecules and be recognized by specific T cell receptors.

Antigen presentation: the ability of an APC to display foreign antigens bound to MHC molecules on its surface and be recognized by specific T cells.

Antigen receptors: the molecules on lymphocytes that bind to antigens or antigen/MHC combinations: antibodies (Ig) on B cells and T cell receptors on T cells.

CD molecules: a system of nomenclature for surface molecules on cells of the immune system, including:
 CD3—a complex of three or more chains, associated with T cell antigen receptor heterodimers, that is involved in antigen specific T cell activation.
 CD4—a molecule expressed by many T helper cells that interacts with MHC class II molecules, mediates cell adhesion, and delivers regulatory signals.
 CD8—one chain of a heterodimer (and possibly a homodimer in humans) expressed by many cytotoxic T cells that interacts with MHC class I molecules and has analogous functions to CD4.
 CD11a, CD11b, CD11c—the α chains of a group of molecules (LFA-1, CR3 and p150,95, respectively) present on several types of leukocyte, and which can mediate adhesion of molecules or other cells.
 CD18—the common β chain of the CD11 family of molecules.

Cytolysins: soluble molecules that are secreted by T_C when they recognize APCs (target cells) and which mediate killing of the latter. They include pore-forming molecules (perforins) and non pore-forming molecules.

Dendritic cells: usually refers to mature 'lymphoid' dendritic cells that can be isolated from lymphoid tissues. These cells are potent immunostimulatory cells.

Dendritic leukocytes: an apparently distinct lineage of leukocytes distributed throughout the body, many of which have or can adopt a characteristic cytology, but which more importantly play a pivotal role in inducing immune responses.

Effector cells: cells that bring about the removal of antigen from the body (e.g. T_C cells and plasma cells).

Epitope: the region of an antigen which binds to an antigen receptor.

Follicular dendritic cells: non-lymphoid cells in B cell areas of lymphoid tissues that retain antigens on their membranes, perhaps in a native form complexed to antibody and complement, and which may be involved in B cell memory. They are thought to be dendritic leukocytes, but this is not proven.

Graft versus host reactions: allogeneic immune reactions caused by donor lymphocytes that recognize histocompatibility molecules on recipient tissues.

Immunostimulation (or sensitization): the activation of small resting lymphocytes into lymphoblasts; the latter are also called activated or sensitized lymphocytes.

Immunostimulatory cells: cells that can activate resting T cells, particularly mature dendritic cells.

Interdigitating cells: non-lymphoid cells in T cell areas of lymphoid tissues which seem closely related to isolated lymphoid dendritic cells. They may be immunostimulatory cells that express processed antigens bound to MHC molecules and which activate resting T cells.

Interleukins: a group of molecules involved in signalling between lymphocytes, accessory cells and other cells in the body.

Langerhans cells: dendritic leukocytes in the epidermis of skin that develop in culture into mature, immunostimulatory dendritic cells.

Marginal zone: in the spleen, an area between the red pulp and white pulp where lymphocytes can leave the blood circulation.

Major histocompatibility complex (MHC): a group of genes encoding cell surface glycoproteins that include class I and class II MHC molecules, which bind foreign antigens and permit recognition by T cells. The MHC also contains class III genes which encode unrelated molecules.

MHC restriction: the requirement that T cells recognize antigens bound to MHC molecules.

Mixed leukocyte reaction (MLR): the proliferative response of alloreactive T cells when they are cultured with allogeneic accessory cells, particularly dendritic cells. Formerly known as the mixed lymphocyte reaction because the stimulating cells were thought to be lymphocytes.

Passenger leukocytes: cells present in a graft that sensitize the recipient against graft antigens, perhaps especially dendritic leukocytes.

Repertoire: the total range of antigen receptor specificities within an individual's immune system.

Self: that environment in which the immune system has developed.

Self-restriction: the phenomenon whereby most T cells preferentially seem to recognize foreign antigens bound to self-MHC molecules; recognition of foreign antigens bound to non-self-MHC molecules is called allorestriction.

TD/TI responses: description of whether immune recognition by T cells is (T-dependent) or is not (T—*in*dependent) required for the generation of an immune response, particularly with reference to antibody responses.

Tolerance: strictly, the failure of lymphocytes to react to an antigen which they have recognized. More generally, the unresponsiveness of lymphocytes to self components.

Ubiquitin pathway: a pathway of intracellular proteolysis, in which proteins destined for degradation are first tagged with the 8.5 kd polypeptide ubiquitin.

Veiled cells: dendritic leukocytes in afferent lymph. Those found in cutaneous lymphatics appear to be transitional in phenotype between Langerhans cells and mature interdigitating cells or lymphoid dendritic cells. Veiled cells are thought to transport antigens, perhaps in processed form, from the periphery to the draining lymph nodes.

Index

Accessory cells, 6, 46, 74 (*see also* Antigen-presenting cells; Immunostimulatory cells)
  functions, 46, 56
  in culture, 28–45
  in vivo, 59–73
Acquired immunodeficiency syndrome (AIDS), 20
Activation,
  of macrophages, 19
  of T cells, *see* Immunostimulation
  signals, 35, 55
Adhesion, 11
Allografts, 3, 6, 22, 74
  cardiac, 63, 66
  of skin, 64, 66
  rejection of, 21–23, 59, 62–64, 66
  renal, 65
Alloreactivity, 13–15, 74
Alymphatic pedicles, 64, 65
Antibody-dependent cellular cytotoxicity (ADCC), 19–20, 26
Antigens, 12, 46
  33D1, 29, 32, 43, 63
  CD2, 3
  CD3, 34
  CD4, 9, 10, 11, 41, 68
  CD8, 9, 10, 68
  CD18, 32
  CD23 (IgE receptor), 24
  CD45, 22, 25, 29, 32
  endogenous, 46, 53
  exogenous, 46, 53
  F4/80, 20, 29, 32, 41, 70
  native, 11, 54–55
  processed, 11, 12, 24, 28
  T-dependent (TD), 23, 34, 76
  T-independent (TI), 23, 76
Antigen-binding sites, 14
Antigen presentation, 3, 17–27, 37, 49, 74
  by B cells, 23–24
  by dendritic leukocytes, 56–57
  by macrophages, 17–21
Antigen-presenting cells, 3, 28, 35–36, 46, 74
  B cells, 18
  macrophages, 18

Antigen processing, 12, 46–58, 74
Apoptosis, 5
Arachidonic acid metabolites, 20
ATXBM mice, 22, 71
B blasts, 36
B cells, 1–3, 18, 23, 34, 49–51
  antigen presentation by, 23
  antigen recognition by, 1
  epitopes for, 11
  responses of, 23, 33
  tolerance of, 71
Bacterial infections, macrophage activation and, 21
Biozzi mice, 20
Birbeck granules, 41, 43
Blastogenesis, 6
Bone marrow, transplantation of, 13, 69, 71

Chloroquine, 49, 50, 51, 53
Class I MHC molecules, 9, 13, 53
Class II (Ia) MHC molecules, 8, 49
Clonal deletion, 71
Clonal expansion, 6
Clustering, 55, 62
Coagulation factors, 19
Coated pits, 47
Complement, 1, 5, 19
  receptor, 2, 18, 20, 39, 43
Concanavalin A, 34
Contact sensitivity, 21, 59
Cytokines, 9, 19
  receptors for, 2
Cytolysins, 3, 5, 6, 18, 75
Cytotoxic lymphocytes (CTL), 18, 33
  antigen recognition by, 51
  cytolysin secretion by, 18
  development of, 18, 22, 34
  responses of, 34
Cytotoxic T cells (T$_C$), 22, 23, 34, 51
  antigen recognition by, 8, 12
  CD8 expression by, 9
  cytolysin production by, 3, 18

Dendritic cells,
  lymphoid, *see* Lymphoid dendritic cells

non-lymphoid, 43
Dendritic leukocytes, 36–44, 75
  antigen presentation by, 56–57
  in transplantation, 62–67
  maturation of, 60–62, 66
  migration of, 60–62
Deoxyguanosine, 63, 70, 71
Determinant selection, 12

Effector cells, 3, 75
Effector mechanisms, 1
Endocytosis, 43, 47–49, 57
  coated pits, 47
  CURL, 48
  lysosomes, 47
  membrane recycling and, 47
Endoplasmic reticulum, 49
Eosinophils, 26
Epitopes, 75
  conformational, 11
  for B cells, 11
  for T cells, 11, 49, 50
  sequential, 11
Exocytosis, 48

Fc receptors, 2, 18, 19, 20, 41
Fibroblasts, 51–55
Follicular dendritic cells, 39, 75

Golgi body, 49, 52
Graft-versus-host responses, 23, 75
Granulocyte–macrophage colony-stimulating factor (GM-CSF), 25, 41, 43, 60

Haptens, 23, 50
Helper T cells, 24–26
  antigen recognition by, 8
  CD4 expression by, 9
  lymphokine secretion by, 3, 4, 18, 20
High endothelial venules (HEV), 62
Histocompatibility locus 2 (H-2), 6, 8
Homing receptors, 62
Host defence, 18
Human immunodeficiency virus (HIV), 20
Human leukocyte antigen (HLA), 6
Hypersensitivity,
  type I (immediate), 11, 25, 26
  type IV (delayed), 11, 21, 25, 26, 60, 61

Ia, see Class II molecules
Immune complexes, 39
Immunoglobulin (Ig), 1, 2, 74
  antigen binding by, 24
  fold, 8
  membrane, 49, 50–51
  responses of, 33, 59
  superfamily, 8
Immunostimulation, 6, 17, 18, 28, 37, 43, 46–58, 61–62, 75

  central, 64, 66
  peripheral, 64, 65, 66
Immunostimulatory cells, 35, 46, 75
  dendritic cells as, 18, 28–35
Inflammatory cells, 1
  phagocytes as, 1
Influenza, 51
Interdigitating cells, 36–39, 65, 68, 75
Interferon-γ, 4, 5, 18, 24, 26
  allogeneic reactions and, 23
  macrophage activation and, 20–21
Interleukins, 5, 18, 75
  IL1, 39, 55
  IL2, 4, 22, 24, 26, 39, 55
    receptor, 24, 26, 32, 41, 68
  IL3, 4, 24, 26
  IL4, 4, 22, 24, 26
  IL5, 4, 24, 26
  IL6, 4, 24, 56
Isotype switching, T cells control of, 23

Langerhans cells, 21, 41–43, 56, 61, 75
  Birbeck granules of, 41, 43
  maturation of, 60
  migration of, 64
Lipopolysaccharides, 24
Lymphocyte,
  activating signals, 55
  recirculation, 62
  repertoire, 1
Lymphoid dendritic cells, 17, 29, 31, 32, 33, 56, 64, 75
  features, 28–29
  follicular, 39, 75
  functions, 32–35, 59–60
  in thymus, 67–72
  migration of, 65–66
Lymphokines, 3–5, 21, 22
  production of, 6
Lysosomes, 47
  proteolysis and, 46
Lysosomotropic agents, 49
  chloroquine, 49, 50, 51, 53

Macrophages, 5, 8, 19, 29, 30, 31, 33, 47–49, 56, 68
  activation of, 17–20
  endocytosis and, 28, 47
  inflammatory (stimulated/elicited), 18, 20
  resident (resting), 17, 20
Major histocompatibility complex (MHC), 2, 49, 52, 75
  class I molecules, 8, 9, 13, 53
  class II (Ia) molecules, 8, 49
  genes, 6, 8, 9
  interactions with peptides, 12–13
  polymorphisms, 13
  restriction, 6–11, 24, 28, 69, 75
Mannosyl–fucosyl receptor, 18, 20

Index

Mast cells, 26
Mechanisms for graft rejection, 21
  sensitization and, 64
Memory cells, 6, 35
Memory responses, 33, 34
β_2-microglobulin, 8, 53
Mitogen responses, 33
Mixed leukocyte reaction (MLR), 32–33, 37, 75
  allogeneic, 14, 33
  syngeneic, 32, 33
MNLX TDL, 41

Necrosis, 1
Non-lymphoid dendritic cells, 43
Nude mice, 21, 69

Opsonization, 2
Oxidative mitogenesis, 34

Passenger leukocytes, 62, 63, 76
Peptide–MHC interactions, 12, 13
Perforin, 5
Phagocytes, 1
Phagocytosis, 47
Pinocytosis, 47
Plasma cells, 1, 5
Polymorphisms, MHC and, 13
Protozoal infections, macrophage activation and, 21
Privileged sites, 64

Reactive oxygen intermediates, 19, 20
Receptors,
  for antigen, 1
  for IgE, 24
  for complement, 18, 20, 39, 43
  for cytokines, 2
  for IL2, 24, 26, 32, 41, 68
  for Fc, 2, 18, 19, 20, 41
  of B cells (Ig), 1, 2, 3, 10
  mannosyl–fucosyl, 18, 20
Restriction repertoire, 69–71
Rough endoplasmic reticulum, 31, 51, 52

Secondary lymphoid tissue, 17
Self-restriction, 13–15, 69, 76
Self-tolerance, 69, 71–72
  of T cells, 17
Sensitization, see Immunostimulation
Signal peptide, 51
Signals, CD4/CD8 and, 11
Spleen, 39, 65–66
Stimulated cells, 18
Supressor T cells, 3

T cell receptors, 1, 2, 74
  $\alpha\beta$, 3
  $\gamma\delta$, 3
  V genes, 10, 51

T cells, 3–6
  activated, response to APCs, 6
  ancillary molecules and, 3
  antigen recognition, 1, 3, 6–15
  cytotoxic (T_C), 3, 8, 18
  development of, in thymus, 67–69, 71
  epitopes for, 11, 49, 50
  prediction of occurrence, 12
  helper (T_H), 3, 8, 18, 24–26
  isotype switching, control of, 23
  lymphokine secretion by, 21
  repertoire of, 1, 17
  resting, 35
  self-tolerance of, 17
  sensitized, 6, 35, 67
  suppressor (T_S), 3
T-dependent (TD) antigens, 23, 34, 76
Tetraparental (allophenic) mice, 22
Thymic-restricting cells, 71
Thymocytes, 10, 68
Thymus, 17, 39, 67, 70
  epithelial cells in, 69
  interdigitating cells in, 71
  nurse cells in, 68
  transplantation of, 63
  T cell development in, 67, 68
T-independent (TI) antigens, 23, 76
Tolerance, 76
  of B cells, 71
  of T cells, 71–72, 76
Transplantation,
  antigens on allografts, T cell recognition of, 3
  cardiac, 63
  of bone marrow, 13
  of Islets of Langerhans, 63
  of skin, 64
  of thymus, 63
  renal, 63
Tumour necrosis factors (TNF), 18
  TNFα, 5, 19, 20, 23
  TNFβ, 5, 26

Ubiquitin pathway, 46, 53, 76
  proteolysis and, 53

Veiled cells, 41, 60, 76
Viral antigens, T cell recognition of, 3, 12